# Collection of Exercises for PLC Programming

# PREFACE

I am happy to have the opportunity to publish a PLC programming book for the third time. This time, it's a collection of exercises for learning PLC programming.

For the past seven years I have been a trainer for PLC control in the Automation Technology program (a two-year higher education program) at Dania Academy in Denmark; I have developed many exercises for the students, and the best of these are now found in this book.

**Basis of knowledge**

The ideas for the exercises have come to me by visiting a wide range of trade fairs, updates on social media, literature on automation, as well as from the students' exam projects. Additionally, I get inspiration from my professional work with programming.

**Selection of exercises**

I have chosen exercises for the book that I believe provide good learning within PLC, and at the same time, exercises that the students find exciting. I have spent a long time developing them to ensure that they cover many of the professional challenges that can arise in PLC programming, as well as ensuring that they cover many professional aspects within PLC programming.

The book is pedagogically and didactically structured with more than 150 illustrations and tables, which means that there is a progression in learning throughout the book, and therefore, the exercises increase in difficulty.

Unfortunately, I am unable to assist with the exercises or provide solutions to them, but comments, critiques, and suggestions for improvements are warmly welcomed via email at: TomMejerAntonsen@gmail.com

I have previously published the following books:

"PLC Control with Structured Text (ST)", first edition 2018

"PLC Control with Ladder Diagram (LD)", first edition 2021

These books are now used in a wide range of educational programs and schools, and many companies have purchased them for their employees. Naturally, tips for some of the exercises can be found in the books.

Thanks to students for review, feedback, and inspiration.

I hope you enjoy the book.

Best regards,

Tom Mejer Antonsen
Randers, Denmark, April 2024

**Tom Mejer Antonsen**

# Collection of Exercises for PLC Programming

**100 programming exercises from beginner to expert level**

*1. Edition, April 2024*



*Illustrations and graphics: Tom Mejer Antonsen*

*The original version (Danish) 1. Edition issued March 2024*

*Publisher: BoD – Books on Demand, Hellerup, Denmark*
*Press: BoD – Books on Demand, Norderstedt, Germany*

*ISBN: 978-87-4305-780-2*

# Table of Contents



 

# 1 Introduction

This book contains exercises for programmable logic controllers (PLC).

It is pedagogically structured, starting with simple programming exercises involving only logic (relay logic with on/off signals). After this, more programming techniques are added with increasing the difficulty. This means that after the simple logic exercises, exercises with counters, timers, analog sensors, applied mathematics, ARRAY, STRUCT and STRING are included. The book also includes exercises where a safety PLC must be programmed, as well exercises with component and sensor naming, programming of functions and function blocks, sequence technique, manual or automatic operation, integration with a robot, frequency converters (VFD), or vision cameras.

The final chapter, chapter 11, consists of mixed complex exercises where all programming techniques are combined. For example, there are exercises for production lines and data collection, as well as ones in which sensors must be selected, and the control specification must be written before programming.

All exercises in the book are selected to be relevant, practical, and realistic in relation to the typical challenges that may arise with PLC programming.

**Getting started with the exercises**

If you are a beginner in PLC programming, it is recommended to start with the first exercises in the book. The initial exercises are at a beginner level and provide a good introduction to programming in a PLC. You can go to the more complex exercises later in the book as you gain more experience.

**Choice of PLC brand and programming language**

The exercises can be solved in all types of PLCs, regardless of brand or model. Similarly, there is no requirement to use a specific PLC programming language. You will gain a lot of learning by solving the tasks in several differing PLC programming languages. For example, an exercise can be programmed first in Ladder Diagram (LD) and then in Structured Text (ST).

Some exercises will naturally be more suitable for a specific programming language, so it is up to the you to choose the right one. Similarly, some exercises are best for using sequence techniques and developing own functions.

**Who can use this book?**

The book is primarily designed for use for the higher education Academy Profession (AP) graduate in Automation Engineering. However, the book can also be used for many other educations that include PLC programming. This could be electrician training, control and regulation electrician training and automation technicians, as well as higher education programs for mechatronics engineers or automation engineers.

Sorry, no guarantee, support, or solutions are provided for the exercises in the book.

# 2 Logical circuit exercises

This chapter contains exercises with logical circuits only; the first exercises are at beginner-level. Later in the chapter, more challenging exercises will be introduced. All tasks can be solved using relay logic (on/off signals).

## 2.1 Turn on light in lamp with a push button

This exercise has a push button switch **S1** and a lamp **P1** connected to a PLC:

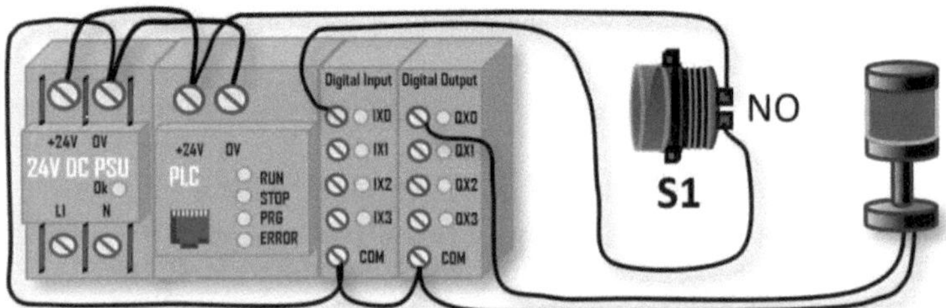


The push button switch **S1** is connected to a digital input module, and the lamp **P1** is connected to a digital output module.
The push button switch is a Normally Open (NO) contact.

Write a PLC program that meets the following requirements:

- The lamp **P1** is turned on when **S1** is pressed.
- The lamp **P1** should be off when **S1** is not pressed.

## 2.2 Turn on light in lamp with two push buttons

Here are two push buttons and a lamp connected to a PLC:

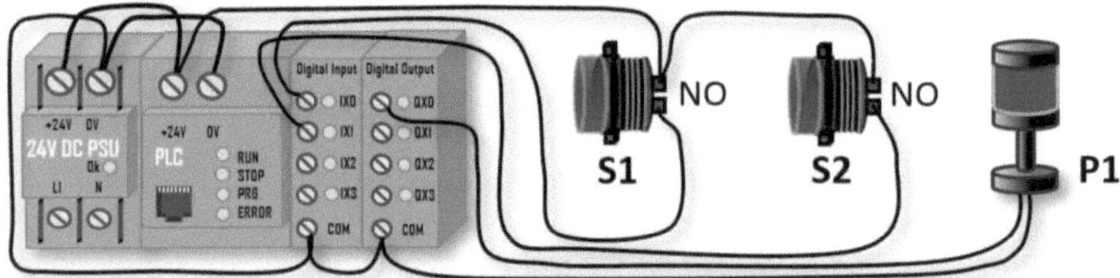


Both push buttons are Normally Open (NO) contacts.
The two push buttons are connected to separate digital input on a digital input module, and the lamp is connected to a digital output module.

Write a PLC program where lamp **P1** is turned on when both push buttons **S1** and **S2** are pressed simultaneously (at the same time). If both push buttons are not pressed simultaneously, the lamp should remain off.

 

## 2.3 Turn off light in lamp with push button

Here's a push button with a lamp connected to a PLC:

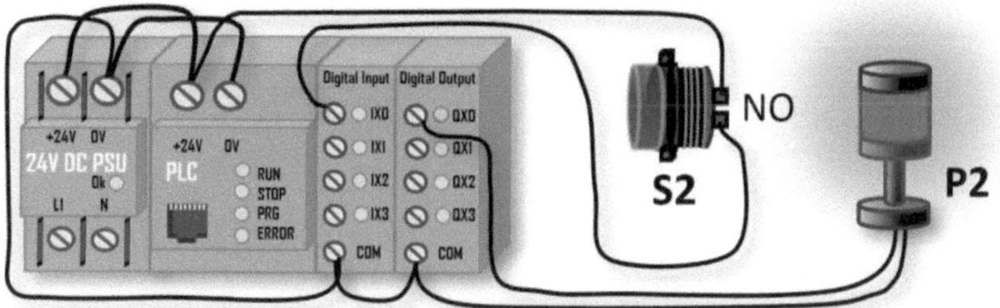


The lamp is turned on.
The push button is a Normally Open (NO) contact.

Write a PLC program that meets these requirements:

- When button **S2** is pressed, the lamp **P2** should be turned off.
- When button **S2** is not pressed, the lamp **P2** should remain on.

## 2.4 Turn on light with one of the two push buttons

This exercise has two push buttons and a lamp connected to a PLC:

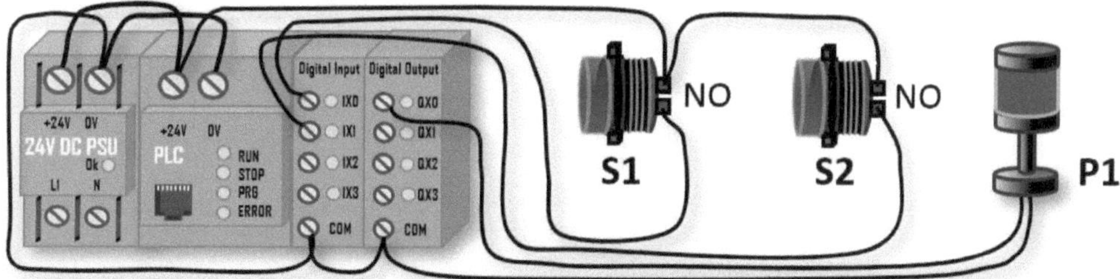


Both push buttons are Normally Open (NO) contacts.

Write a PLC program where the lamp **P1** is turned on when either push button **S1** or push button **S2** is pressed.

## 2.5 Turn on two lamps with a rotary switch

In this exercise, a rotary switch is connected to a digital input, and two lamps are connected to separate digital outputs:

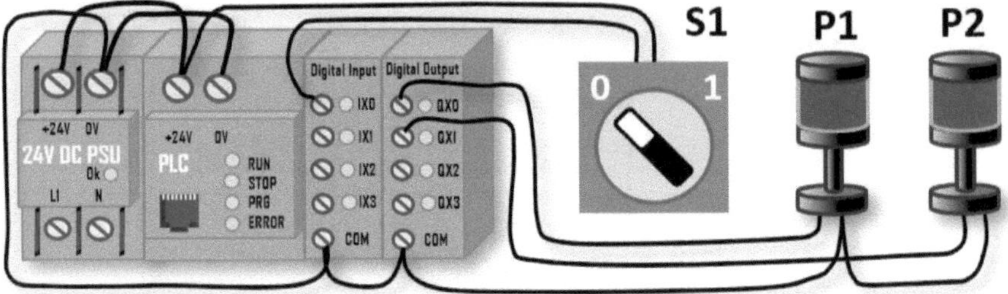


Write a PLC program that operates as follows:

- If the rotary switch **S1** is in position "1" (on), both lamps should be turned on.
- If the rotary switch **S1** is in position "0" (off), both lamps should be turned off.

## 2.6 Turn on and off lamps with a push button

This exercise has a push button and two lamps connected to a PLC:

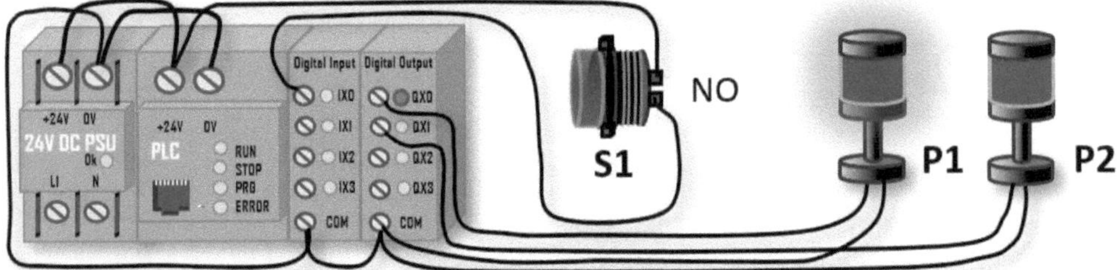


One of the lamps is turned on.

Write a program that operates as follows:

- When push button **S1** is pressed, the lamp **P1** should be turned off, and the lamp **P2** should be turned on.
- When push button **S1** is not pressed, the lamp **P1** should be on, and the lamp **P2** should be off.

 

## 2.7 Turn on lamp with two push buttons (NO, NC)

Here are two push buttons, **S1** and **S2**, and a lamp, **P1**, connected to a PLC:

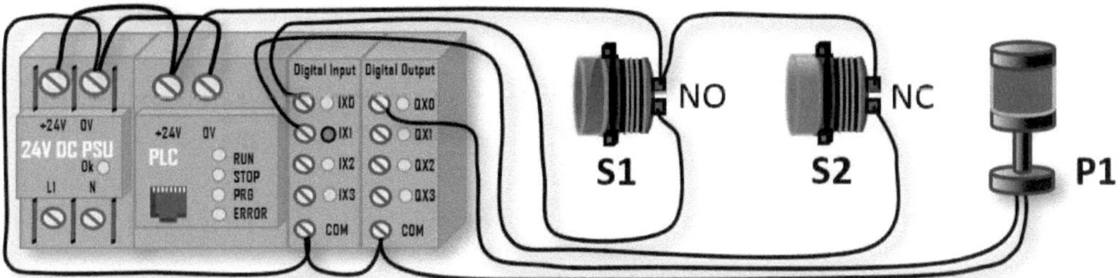


Push button **S1** is a Normally Open (NO) contact.
Push button **S2** is a Normally Closed (NC) contact.

Write a PLC program that operates as follows:

- The lamp **P1** should be turned on if either push button **S1** or **S2** is pressed.
- If both push buttons **S1** and **S2** are pressed at the same time, the lamp **P1** should also be turned on.

## 2.8 Turn on lamp tower with two push buttons

Here are two push buttons and a lamp tower with three lamps connected to a PLC:

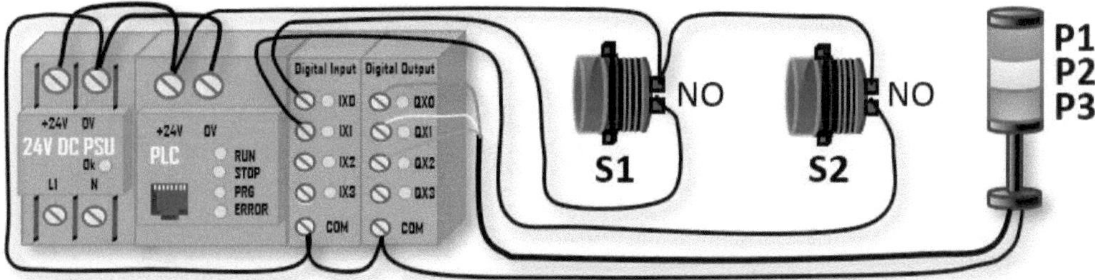


Write a PLC program that operates as follows:

- When button **S1** is pressed, the lamp **P1** should be turned on.
- When button **S2** is pressed, the lamp **P2** should be turned on.
- When both buttons are pressed at the same time, **P1** and **P2** should be turned off.
- When both buttons are pressed at the same time, **P3** should be turned on.

## 2.9 Turn on lamp tower with a rotary switch

This exercise has a manual rotary switch with four settings: **S0**, **S1**, **S2**, and **S3**; each is connected to a separate digital input on the PLC. Further, there's a lamp tower with three lamps connected to digital outputs:

Write a PLC program that operates as follows:

The setting of the rotary switch must control the lighting in the lamp tower:

- Setting 0: All three lamps should be off (as shown in the image).
- Setting 1: Only lamp **P1** should be turned on.
- Setting 2: Only lamp **P2** should be turned on.
- Setting 3: All three lamps should be turned on.

## 2.10 Programming of self-hold

In this exercise, two push buttons and a lamp are connected to a PLC:

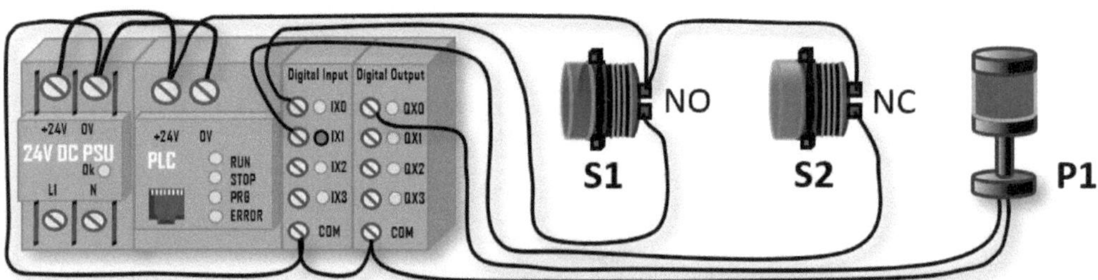


Push button **S1** is a Normally Open (NO) contact.
Push button **S2** is a Normally Closed (NC) contact.

Write a PLC program that operates as follows:
The lamp **P1** should turn on when the push button **S1** is pressed, and the lamp should remain turned on even if push button **S1** is no longer pressed. When the push button **S2** is pressed, lamp **P1** should turn off.
This is a fundamental function in a PLC known as self-hold.

 

## 2.11 Package on conveyor belt

In this exercise, you must write a PLC program to control a conveyor belt.

Illustration of the conveyer belt:

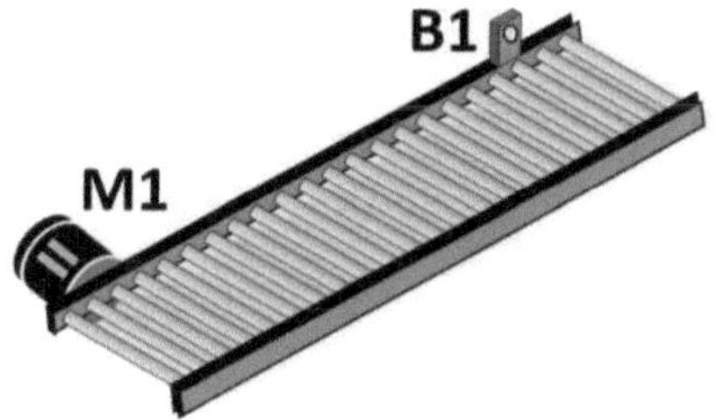


**Description**

The setup consists of a conveyor belt controlled by a motor named **M1**. The motor runs when it receives a **TRUE** digital signal from the PLC and stops when the digital signal is **FALSE**.

The conveyor belt has a sensor **B1** that gives a **TRUE** signal when there is a package at front of the sensor. When there is no package, the sensor gives a **FALSE** signal.

There is a rotary switch **S1** used to start and stop the conveyor belt.

The two operating situations are shown below:

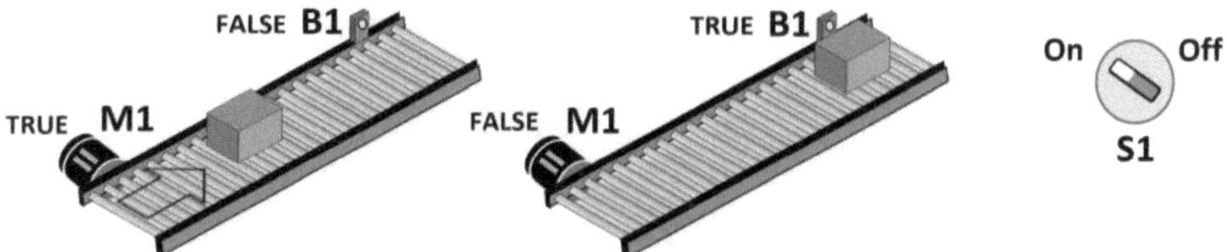


**Operation**

The rotary switch **S1** is used to turn on (run) and off (stop) the conveyor belt.

Packages come onto the conveyor belt, and when sensor **B1** is activated (gives a **TRUE** signal) by a package, the conveyor belt must stop to prevent the package from falling off the conveyor belt and ending up on the floor.

Once an operator has removed the package stopped at sensor **B1**, the belt should start again to allow for a new package to come in.

**Exercise**

Write a PLC program based on the description.

 

## 2.12 Pallet lift table

In this exercise, you must write a program for a pallet lift table:

**Purpose**

The pallet lift table can raise a pallet to working height, allowing employees to access the goods on the pallet without needing to bend down. This solution provides a good working environment.

**Description**

There is an actuator **M1** (a motor cylinder) that operates the pallet lift up and down.

The pallet lift is operated by a foot panel with two switches. The lift moves up when switch **S2** is held down and moves down when switch **S1** is held down. A sensor **B2** gives a **TRUE** signal when the pallet lift should not move higher up, and a sensor **B1** gives a **TRUE** signal when the pallet lift should not move further down. The pallet lift must not move if both switch **S1** and switch **S2** are pressed down at the same time.

Here is a pallet with two packages placed on the pallet lift table:

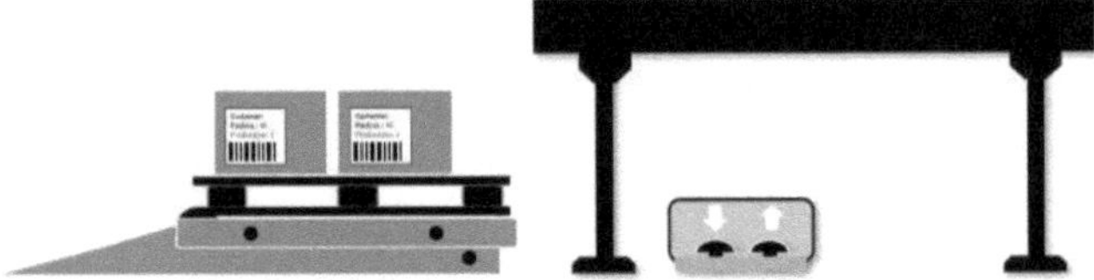

Here, the lifting table has been raised all the way to the workbench:

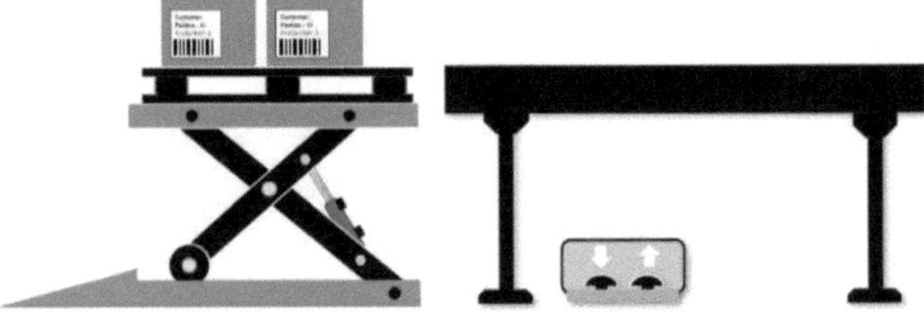

**Exercise**

Write a PLC program for the pallet lift table.

 

## 2.13 Filling water into jerrycans

In this exercise, you must write a PLC program that can fill water into jerrycans.

**Description**

A motor **M1** drives a conveyor belt. When the motor receives a **TRUE** signal from the PLC, the conveyor belt moves to the right.

When sensor **B1** gives a **FALSE** signal and sensor **B2** gives a **TRUE** signal, the jerrycan is positioned correctly for water filling.

The jerrycan is filled with water when valve **Q1** (an electric water valve) receives a **TRUE** signal from the PLC.

Sensor **B5** measures the amount of water filled in the jerrycan. When sensor **B5** gives a **TRUE** signal, the jerrycan is filled, and the valve **Q1** must be set to **FALSE** to shut off the water.

Once the jerrycan is filled, motor **M1** is set to **TRUE** to move the jerrycan away and bring in a new empty jerrycan.

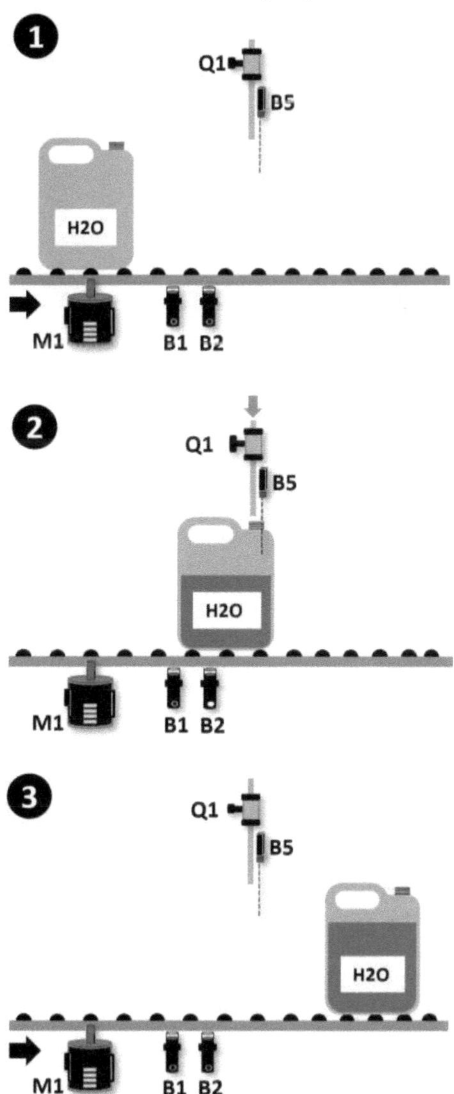


**Description for the images:**

| | |
|---|---|
| 1 | An empty jerrycan moving along the conveyor belt. |
| 2 | An empty jerrycan being filled with water. |
| 3 | The jerrycan is now filled with water and moving away. |

**Exercise**

Write a PLC program based on the description.

## 2.14 Pump well with one pump and two float switches

In this exercise, you must write a program for a pump well.

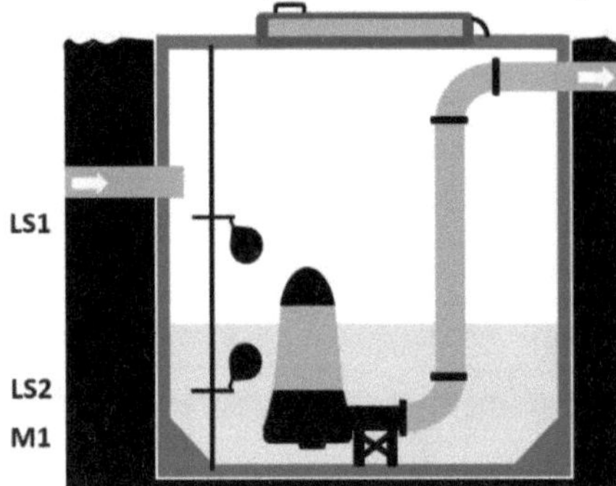


The pump well is located below ground level and consists of a pump and two float switch contacts, as shown in the image to the left. The pump well is used to pump wastewater away from a residential house. The pump is controlled by a motor **M1** and can be directly controlled with an on/off signal (digital signal) from a PLC. The top float switch contact **LS1** is a Normally Closed (NC) contact. The bottom float switch contact **LS2** is a Normally Open (NO) contact.

When wastewater flows into the pump well, the water level in the pump well rises, activating the two float switch contacts as shown below:

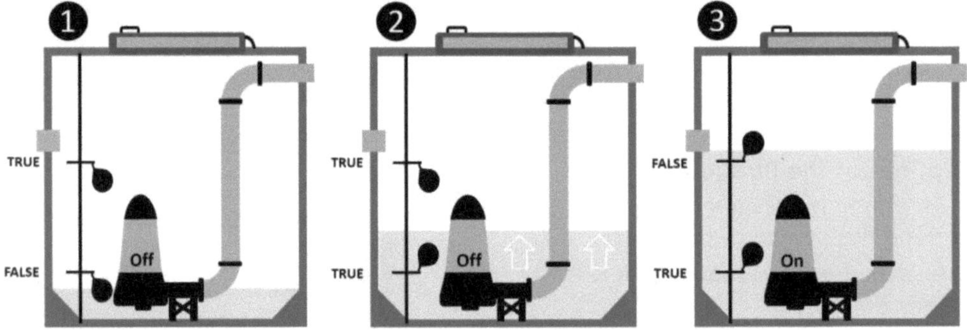


**Low level in the pump well**

When there is a low level (1) in the pump well, float switch **LS2** gives a **FALSE** signal, and the pump **M1** should be turned off since there is no need to pump wastewater.

**Increasing level**

As the level in the pump well rises (2), float switch **LS2** is activated first, followed by float switch **LS1**. It is only when **LS1** is activated (gives a **FALSE** signal) by the level that pump **M1** should start to pump the wastewater away (3).

**Decreasing level**

The pump should continue to run even if the level falls below **LS1**. It is only when the level falls below float switch **LS2** again that the pump should stop.

**Exercise**

Write a PLC program for the pump well.

 

## 2.15 Program for a for cooker hood

In this exercise, you must write a PLC program for a cooker hood used in a kitchen.

**Description**

The cooker hood has a fan inside that can operate at three speeds.
There is built-in lighting in the cooker hood that can be turned on and off.

The cooker hood has a total of six control buttons at the front:

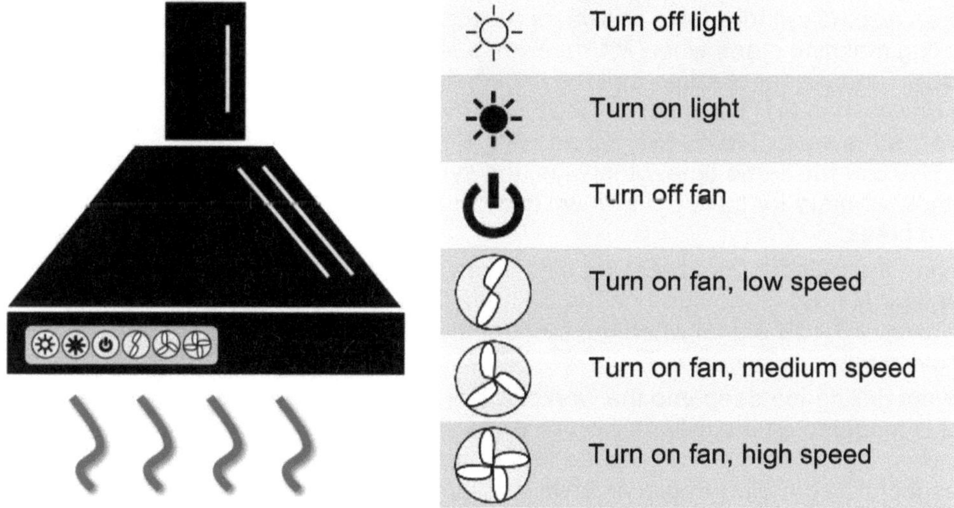


In this exercise, you must choose names for variables and components yourself.

It is not necessary to turn off the fan before changing the speed.

The fan speed is controlled by three digital signals from the PLC, and only one of these signals may be **TRUE** (24V) at any given time. If all three digital signals to the fan are **FALSE** (0V), the fan is turned off.

**Exercise**

Write a PLC program for the cooker hood.

## 2.16 Drilling machine with two-hand controls

This exercise is to write a PLC program for drilling machine with two-hand controls.

The setup consists of a drilling machine **M1**, two manually operated push buttons **S1** and **S2**, and a sensor **B3**.

**Description**

The drilling machine **M1** can move up and down with the help of an electric cylinder (also known as an actuator). The drilling machine starts when **M1** receives a signal (set to **TRUE**). The cylinder ensures that the drilling machine moves down when **M1_K2** is set to **TRUE** and moves up when **M1_K1** is set to **TRUE**. **M1_K2** and **M1_K1** must not be **TRUE** at the same time, otherwise, the cylinder won't know whether to move up or down and could therefore break.

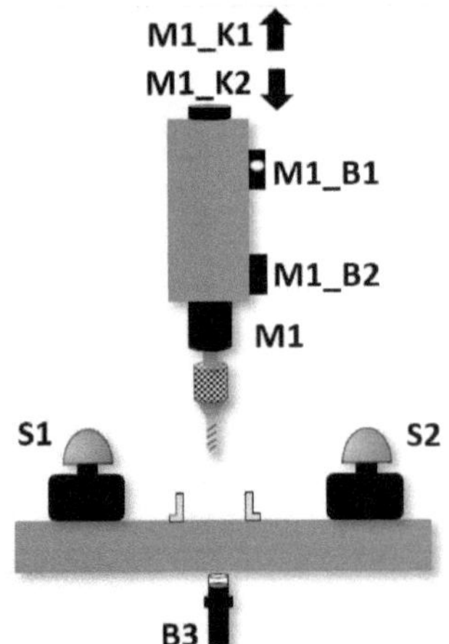


To prevent the cylinder from breaking if it moves too far up, a sensor **M1_B1** is mounted on the cylinder. This sensor gives a **TRUE** signal when the cylinder should stop moving up.

To prevent drilling too deep into the workpiece, a sensor **M1_B2** is mounted on the cylinder, which gives a **TRUE** signal when drilling should not proceed further down. This ensures that holes in all workpieces have the same depth, thus maintaining consistent quality.

The following images show the operation of the drilling machine:

**Workpiece placement**

First, a workpiece must be placed in the holder (fixture) under the drilling machine. When the workpiece is placed, sensor **B3** will be activated (giving a **TRUE** signal).
When sensor **B3** is activated, a small LED (lamp) on the sensor will turn on, indicating to the operator that the workpiece is positioned correctly, and the drilling machine can be started.
In the PLC program you are writing, ensure that the drilling machine can only start if there is a workpiece in place. For this, you can use the signal from sensor **B3**.

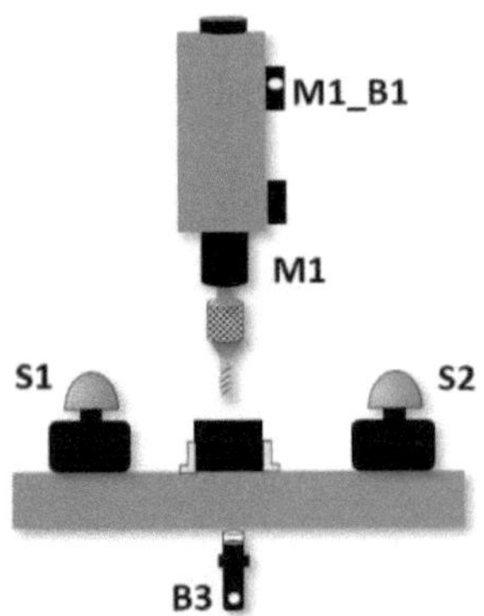


 

**Start drilling machine**

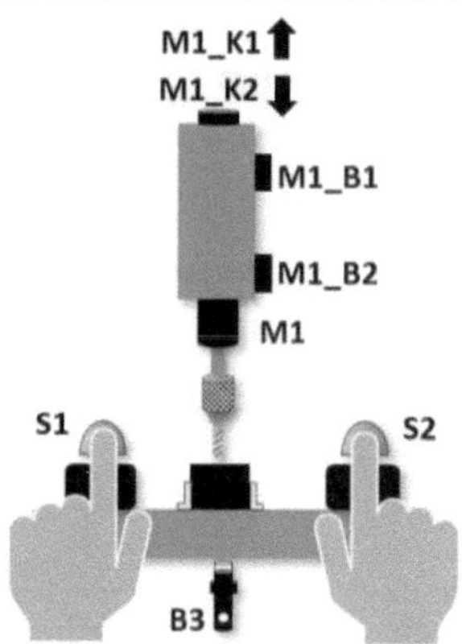


When the workpiece is positioned correctly in the fixture, the operator should start the drilling machine. This is done by keeping one finger on each of the two push buttons, **S1** and **S2**.

Both push buttons, **S1** and **S2**, must be pressed down for the drilling machine to start. This is to ensure that the operator, who operates the drilling machine, does not have fingers close to the drilling machine and thus could be injured when drilling into a workpiece.

When writing your PLC program, you must ensure that the drilling machine can only start if there is a **TRUE** signal from both **S1** and **S2**, as well as **B3**.

**Drilling hole**

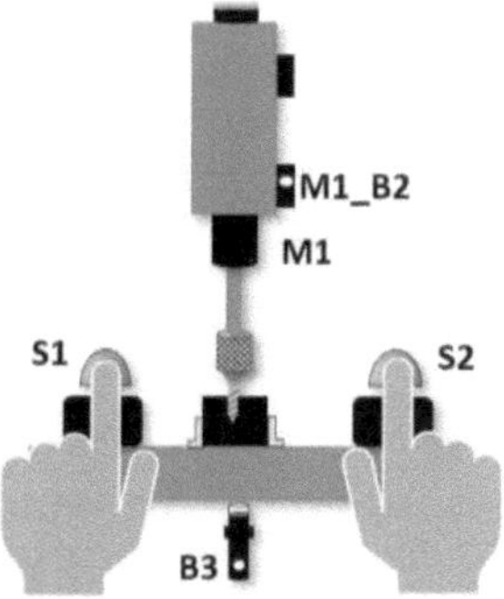


When the light is turned on in the sensor **M1_B2**, the hole is drilled deep enough, and the cylinder should now move up again.

**Final**

When there is a **TRUE** signal on **M1_K1**, the cylinder moves up. The operator must still keep both fingers on the two push buttons. If the operator removes one or both fingers, the drilling machine and the movement of the cylinder must immediately stop to prevent the operator from getting injured.

When there is a **TRUE** signal from sensor **M1_B1**, the drilling machine and cylinder must stop moving. The operator can then remove the workpiece that has been drilled and insert a new workpiece.

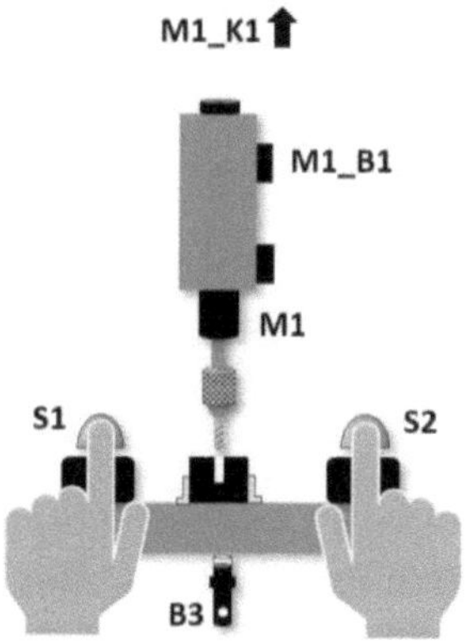

**Exercise**

Write a PLC program for the drilling machine.

## 2.17 Palletizing jerrycans with a robot

In this exercise, you must write a PLC program for a small robotic system that places two jerrycans on a pallet. Below are four images showing the process:

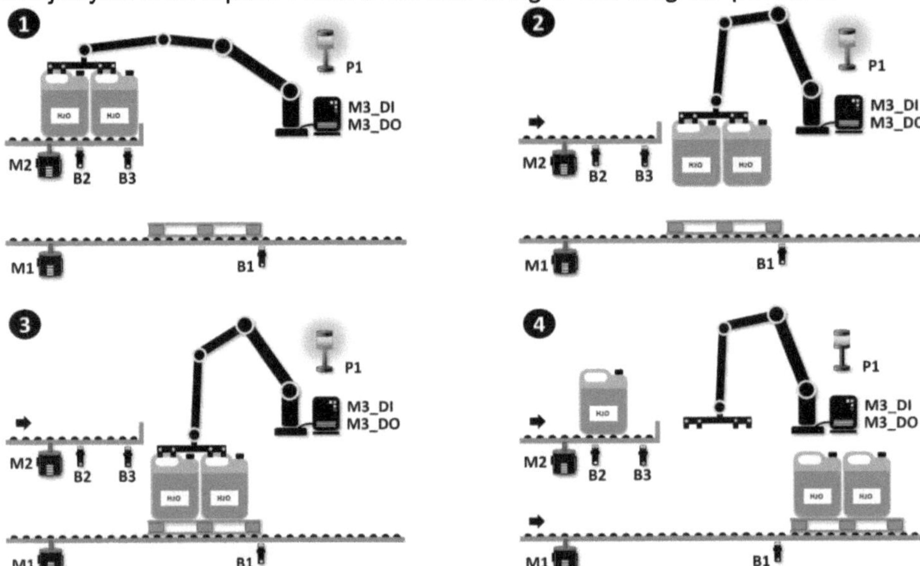


**Description of the images**

The first image (1) shows that the robot has just grabbed the two jerrycans from the upper conveyor belt. Image (2) shows that the robot is moving the two jerrycans down onto the pallet. In image (3), the robot has just placed the two jerrycans on the pallet. The last image (4) shows that the pallet with the two jerrycans is being driven away on the conveyer belt, and the robot is waiting while another jerrycan enters the upper conveyor belt.

**Description**

It is only when there is a **TRUE** signal from sensor **B1**, **B2**, and **B3** that the robot is allowed to begin its work. The robot starts when a **TRUE** signal is set to **M3_DI**, which is a digital input signal to the robot controller. When the robot is working, the output on the robot controller **M3_DO** is set to **TRUE**. When the robot has finished placing the two jerrycans on the pallet and the robot arm has moved away, as shown in image (4), the digital output on the robot controller **M3_DO** is set to **FALSE**.

When the robot is working, the PLC program must turn the lamp **P1** on.

A new empty pallet is moved forward when motor **M1** has a **TRUE** signal. The pallet must be is moved forward until a **TRUE** signal comes from sensor **B1**. Then the pallet is placed correctly for the robot to place the two jerrycans. When the robot has placed the two jerrycans on the pallet, set **M1** to **TRUE** to move the filled pallet away.

**Exercise**

Write a PLC program for the small robotic system.

 

## 2.18 Conveyor for pallet pickup (manual/auto)

In this exercise, you must write a PLC program for a conveyor belt where pallets loaded with goods can be picked up by a forklift.

Illustration:

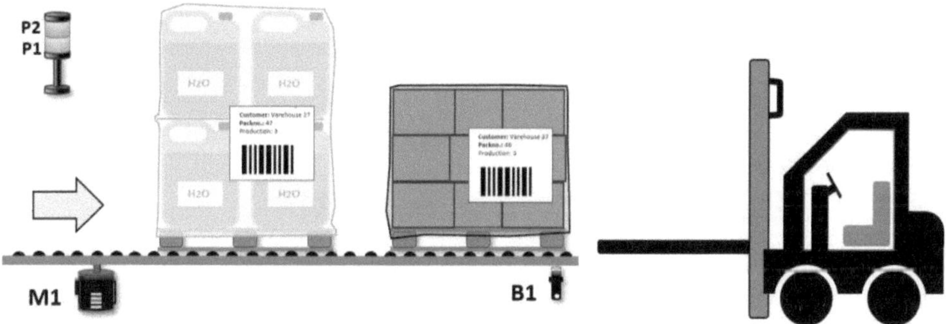


**Description**

The conveyor belt is controlled by the motor **M1**.
A light tower with two lamps, **P1** and **P2**, is used to signal the operator sitting in the forklift. If the lamp **P2** is on, a pallet is ready for pickup on the conveyor belt. If the lamp **P1** is on, the conveyor belt is moving the next pallet to the end of the conveyor belt, where the it is to be picked up. When the pallet is at the end of the conveyor belt, the sensor **B1** gives a **TRUE** signal, at which the conveyor belt should stop. The conveyor belt should run when sensor **B1** gives a **FALSE** signal.

**User interface**

There is a user interface as shown below:

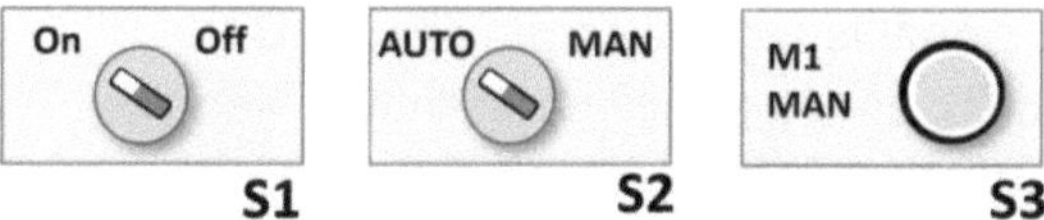


The conveyor belt is set into operation with the manual turn switch, **S1**.
The turn switch **S2** is used to switch between automatic and manual operation mode. If switched to manual operation, the motor **M1** will run when pressing push button switch **S3**.

In both automatic and manual operation modes, the conveyor belt should stop when sensor **B1** gives a **TRUE** signal to prevent the pallet from ending up on the floor.

**Exercise**

Write a PLC program based on the description.

## 2.19 PLC program for roller shutter

The exercise is to write a PLC program for a roller shutter:

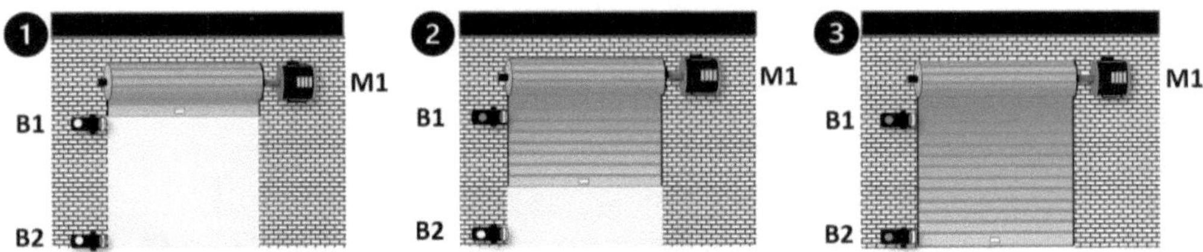


Description of the images:

| Image | Description |
|---|---|
| 1 | Roller shutter fully raised (open). |
| 2 | Roller shutter halfway down. |
| 3 | Roller shutter fully lowered (closed). |

**Description**

The roller shutter is moved by a motor **M1**, which can raise and lower it.
The motor is controlled by two digital signals from the PLC:

**M1Run** turns on the motor with a **TRUE** signal, and
**M1Direction** controls the direction of the motor.

If **M1Direction** is **TRUE**, the roller shutter moves up, and if **M1Direction** is **FALSE**, the roller shutter moves down. To prevent the roller shutter from lowering too far, a sensor **B2** is installed, which sends a **FALSE** signal when it should not move further down. Motor **M1** must stop when a **FALSE** signal is received from **B2**.

To prevent the roller shutter from rising too far, a sensor **B1** is installed, which gives a **FALSE** signal when the roller shutter should not move further up. Motor **M1** should stop when a **FALSE** signal is received from sensor **B1**.

Both sensors **B1** and **B2** are Normally Closed (NC) sensors. This means that the roller shutter will not move too far down or up if a sensor or sensor cable is broken.

**Control Panel**

The roller shutter has this control panel:

Pressing **S1** will raise the roller shutter (move up).

Pressing **S3** will immediately stop moving the roller shutter. After pressing **S3**, either **S1** or **S2** can be used to subsequently raise or lower the roller shutter.

Pressing **S2** will lower the roller shutter (move down).

**Exercise**

Write a PLC program for the roller shutter.

 

## 2.20 Bottle on conveyor belt

In this exercise, you must write a PLC program that can make a bottle move forward and back on a conveyor belt.
The setup consists of an operator panel with a push button **S1**, a motor **M1**, and two sensors **B1** and **B2**. The two sensors are mounted at each end of the conveyor belt. Motor **M1** drives the conveyor belt.
The push button **S1** is a Normally Open contact (NO).
Below are images of the bottle moving forth and back:

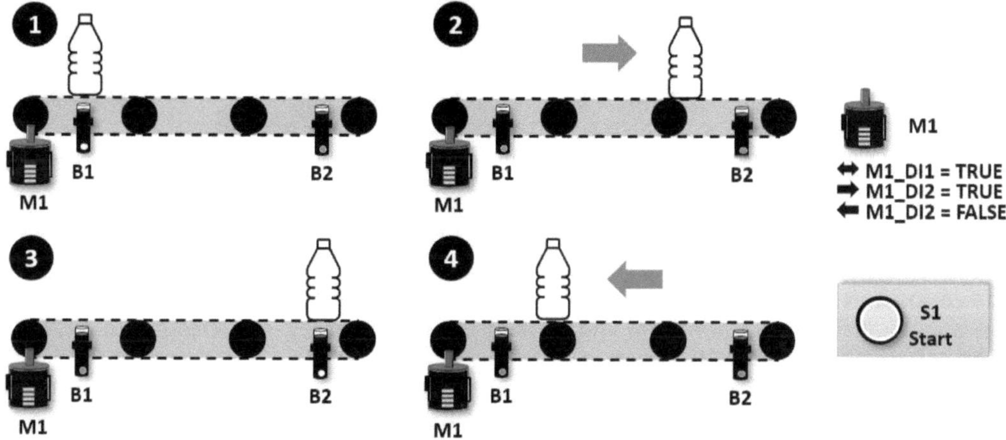


**Description**

The sensors give a **TRUE** signal when the bottle is directly above the sensor.

The motor **M1** is controlled by two digital input signals: **M1_DI1** and **M1_DI2**.

When the signal **M1_DI** is **TRUE**, the motor is in operation and the conveyor belt is moving. The motor can change direction of rotation with the digital signal **M1_DI2**. If **M1_DI2** is set to **TRUE**, the bottle moves to the right, and if the signal is set to **FALSE**, the bottle moves to the left.

**Description of the images:**

| Image | Description |
|---|---|
| 1 | The bottle is placed directly above sensor **B1**, and the push button **S1** can be pressed to start the program. |
| 2 | The program has started, and the bottle is moving to the right. |
| 3 | The bottle has reached sensor **B2**, which can be seen as sensor **B2** gives a **TRUE** signal. Now the conveyor belt needs to run in the opposite direction. |
| 4 | The bottle is on its way back. When the bottle returns to sensor **B1**, the conveyor belt should stop. |

**Exercise**

Write a PLC program based on the description.

 

## 2.21 Production line with jerrycans (manual/auto)

In this exercise, you must write a PLC program for a jerrycan production line:

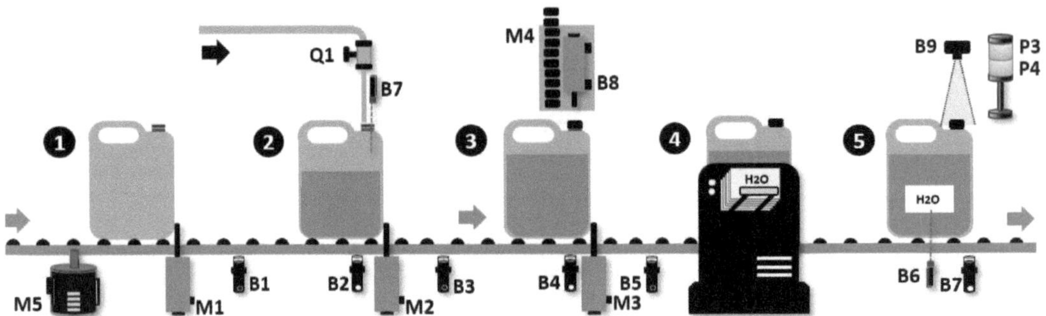


**Description**

The empty jerrycans enter from the right (1). The jerrycan is filled with liquid (2). A cap is put on the jerrycan (3) and a label is applied to the jerrycan (4). Then, the jerrycan is inspected (5).

**Conveyor belt**

The conveyor belt is in operation when the motor **M5** receives a **TRUE** signal and stops when **M5** receives a **FALSE** signal.

**Waiting position for empty jerrycan (1)**

The next jerrycan to be filled waits in front of cylinder **M1**. To allow the empty jerrycan to pass, cylinder **M1** must receive a **TRUE** signal, causing the cylinder to move down. Cylinder **M1** should move up on a falling edge (**TRUE** to **FALSE**) from sensor **B1**.

**Liquid filling (2)**

Cylinder **M2** should be up for the jerrycan to be correctly positioned under the filling nozzle. There must be a **TRUE** signal from sensor **B2** to allow liquid filling into the jerrycan. Cylinder **M2** is up when **M2** receives a **FALSE** signal. Liquid is poured into the jerrycan when valve **Q1** (electric water valve) receives a **TRUE** signal. Liquid pouring should continue until sensor **B7** gives a **TRUE** signal.

When the jerrycan is filled, cylinder **M2** should receive a **TRUE** signal for the cylinder to move down, allowing the jerrycan to proceed. On a falling edge from sensor **B3,** the jerrycan has moved on, and **M2** should receive a **FALSE** signal to move up.

**Capping machine (3)**

Cylinder **M3** should be up, and once there is a **TRUE** signal from sensor **B4**, the jerrycan is in the correct position to receive a cap. The cap is put on by setting **M4** to **TRUE**. The cap is secured and locked on the jerrycan when **B8** gives a **TRUE** signal. Then, cylinder **M3** should be set to **TRUE** to move down, allowing the jerrycan to pass.

On a falling edge from sensor **B5**, the jerrycan has passed the cylinder, and cylinder **M3** should be set to **FALSE** to move the cylinder up again.

 

**Labelling machine (4)**

The machine that applies labels to the jerrycans is a standalone machine with its own PLC control system. The machine independently verifies that the jerrycan is correctly positioned to receive a label. There is no connection or communication between the labelling machine and the PLC program you are writing.

**Cap and label inspection (5)**

When sensor **B7** receives a **TRUE** signal, the jerrycan is ready to be inspected. The jerrycan is considered "OK" when the following conditions are met:

- There is a **TRUE** signal from vision camera **B9**.
- Sensor **B6** gives a **TRUE** signal if the jerrycan has a label.

If the jerrycan is "OK," the lamp **P3** must be set to **TRUE**. If the jerrycan is not "OK," the lamp **P4** must be set **TRUE** to notify the operator that the jerrycan needs to be removed from the conveyor belt.

**Operator panel (HMI)**

The production line is equipped with the following operator panel:

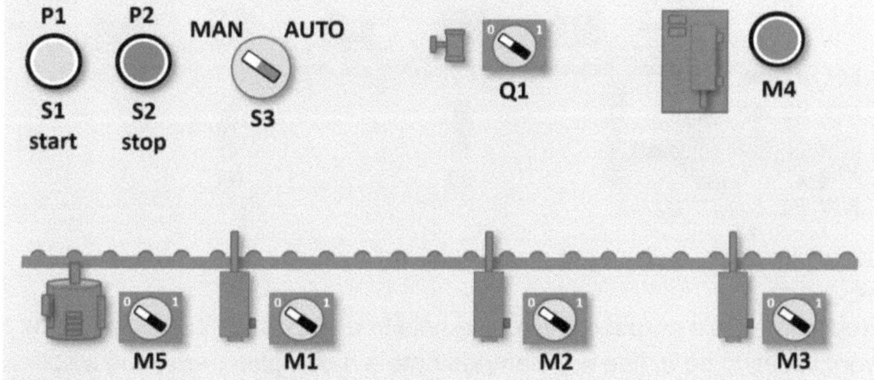


**Start and stop**

Push button **S1** starts the production line and turns on the button light **P1**.
Push button **S2** stops the production line and turns on the button light **P2**.
**P1** and **P2** may not be turned on at the same time.
Rotary switch **S3** is used to switch between automatic and manual operation.

**Automatic operation**

The production line fills liquid into jerrycans automatically, as described earlier.

**Manual operation**

Components can be activated individually with a rotary switch or push button.

**Exercise**

Write a PLC program for the production line.

## 2.22 Production line with two drilling machines

In this exercise, you must develop a PLC program for a small production line with two drilling machines, **M1** and **M2**:

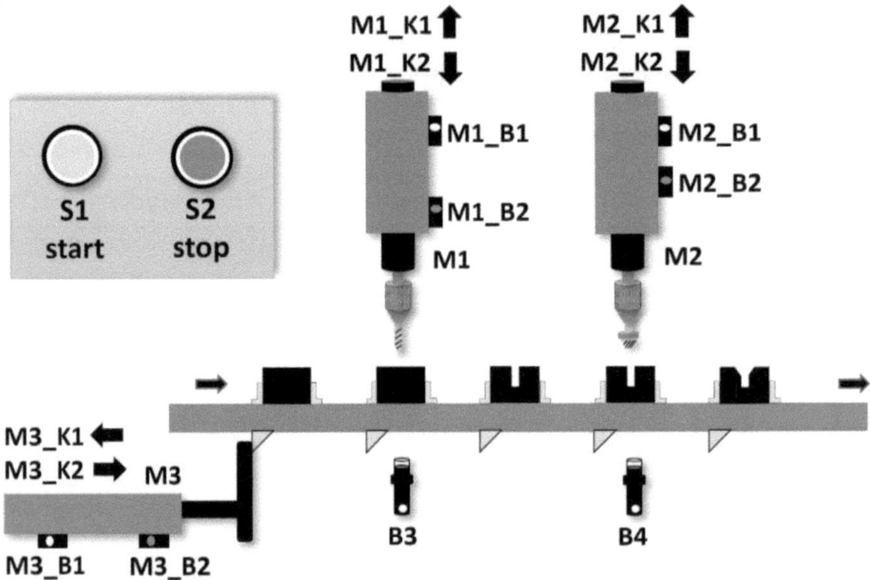


Production line task: drilling and countersinking.

**Description**

The production line's purpose is to drill a hole in a workpiece and countersink it.

The workpieces to be drilled are clamped onto a metal plate, ensuring stability during drilling and countersinking. The metal plate is pushed to the right using cylinder **M3**. When there is a **TRUE** signal on **M3_K2,** the motor in the cylinder pushes the metal plate to the right, advancing the workpieces further. Afterwards, there should be a **TRUE** signal on **M3_K1** to move the cylinder back for the next push.

There are two feedback sensors (reed contacts) on the cylinder:

- **M3_B1** gives a **TRUE** signal when cylinder **M3** is fully to the left (see image).
- **M3_B2** gives a **TRUE** signal when cylinder **M3** is fully to the right.

The **TRUE** signal to **M3_K1** should be set to **FALSE** when a **TRUE** signal is received from sensor **M3_B1**, as the cylinder cannot move further to the left and may malfunction if **TRUE** signal is still on **M3_K1**.

Similarly, the **TRUE** signal to **M3_K2** should be set to **FALSE** when the cylinder is fully to the right and a **TRUE** signal is received from **M3_B2**.

There should not be a **TRUE** signal to **M3_K1** and **M3_K2** at the same time.

 

**The drilling machines**

The two drilling machines operate in the same manner:

Drilling machine **M1** may start when there is a **TRUE** signal from sensor **B3**. The drilling machine operates when there is a **TRUE** signal on **M1**. When there is a **TRUE** signal on **M1_K2**, the cylinder moves the drilling machine lower, allowing a hole to be drilled in the workpiece. When a **TRUE** signal is received from sensor **M1_B2**, the hole is drilled. Then the signal to **M1_K2** must be set to **FALSE**, stopping the drilling machine from lowering further. Afterwards, a **TRUE** signal must be set on **M1_K1** to raise the drilling machine.

Drilling machine **M2** starts when there is a **TRUE** signal from sensor **B4**. The drilling machine operates when there is a **TRUE** signal on **M2**. When there is a **TRUE** signal on **M2_K2**, the cylinder moves the drilling machine lower, allowing the hole to be countersunk. Then the signal to **M2_K2** must be set to **FALSE**, stopping the drilling machine from lowering further. Afterwards, a **TRUE** signal must be set on **M2_K1** to raise the drilling machine.

Once both drilling machines have finished drilling, the metal plate should be pushed to the left to prepare for new workpieces.

Note that both drilling machines do not operate for the same duration, as drilling the hole requires more time than countersinking the hole. The depth to which the drilling machines should lower is physically adjusted with sensors **M1_B2** and **M2_B2**, which can be moved up or down on the side of the cylinder. One drilling machine may remain stationary while the other is in operation. This typically occurs during startup, where **M2** has not yet received a workpiece while **M1** is already drilling into one.

**Operator panel**

As seen in the image, there is a small operator panel with a start switch contact **S1** and a stop switch contact **S2**. This is intended for the operator to start and stop the operation of the production line. If start switch **S1** is activated (pressed), the production line should start. If stop switch **S2** is activated, both drilling machines should stop, and all three cylinders should return to the start position. A cylinder is in a start position when sensor **B1** has a **TRUE** signal as shown in the image.
Both **S1** and **S2** are push-button switches with a spring return function.
**S1** is a Normally Open (NO) contact and **S2** is a Normally Closed (NC) contact.

**Exercise**

Write a PLC program based on the description.

## 2.23 Sorting of packages by size

In this exercise, you must write a PLC program for a small system that can sort packages onto three separate conveyor belts.

Illustration of the system:

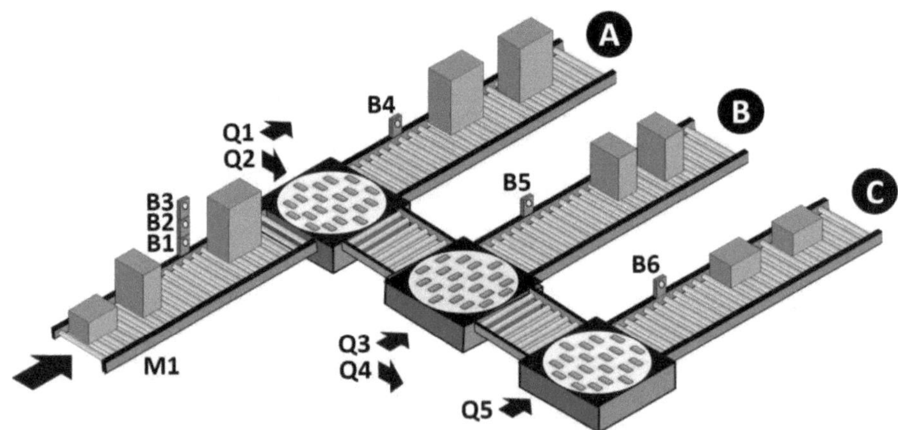


**Description**

Packages enter on conveyor belt **M1**. There are three sizes of packages. The large packages should go to conveyor belt A, medium-sized packages should go to conveyor belt B, and small packages should go to conveyor belt C.
Packages enter conveyor belt **M1** with significant spacing, and all conveyor belts operate (running) continuously.

The sensors **B1**, **B2**, and **B3** are used to determine the size of the package. If there is a **TRUE** signal from all three sensors **B1**, **B2**, and **B3**, it indicates a large package.

The system has three turntable conveyors with small rollers, which can send a package straight or to the side. The turntable conveyors are controlled from the PLC with a digital signal **Q**. This means that if, for example, a **TRUE** signal is set on **Q1**, the large packages are sent out on conveyor belt A, as shown in the illustration. If a **TRUE** signal is set on **Q2**, the package is sent to the side. There should not be a **TRUE** signal on both **Q1** and **Q2** at the same time. Conveyor belts **Q3** and **Q4** function in the same way as conveyor belt **Q1**.
When there is a **TRUE** signal on either **B4**, **B5**, or **B6**, the package has been received on the correct conveyor belt, and the system is ready for a new package.

**Exercise**

Write a PLC program based on the description.

 

## 2.24 Robot and CNC machine controlled by PLC

In this exercise, you must develop a PLC program for a small machine system capable of processing metal workpieces.

The system consists of two conveyor belts, a robot, and a CNC machine, as shown:

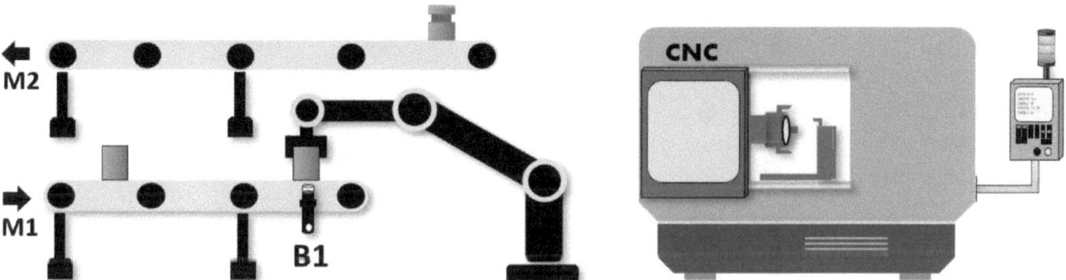


**Description**
The workpieces to be processed come on a conveyor belt **M1**. The robot picks up the workpiece from conveyor belt **M1** and places it in the CNC machine. Once the workpiece has been processed in the CNC machine, the robot removes the workpiece from the CNC machine and places it on conveyor belt **M2**.

**PLC control**
It is a PLC that controls the entire machine system, determining when the CNC machine and the robot should start and stop. This means that there should be electrical connections from the PLC to both the CNC machine and the robot controller, allowing the PLC to start the CNC machine and the robot and to receive a signal back when they are finished.
Cable connections are shown in the image below:

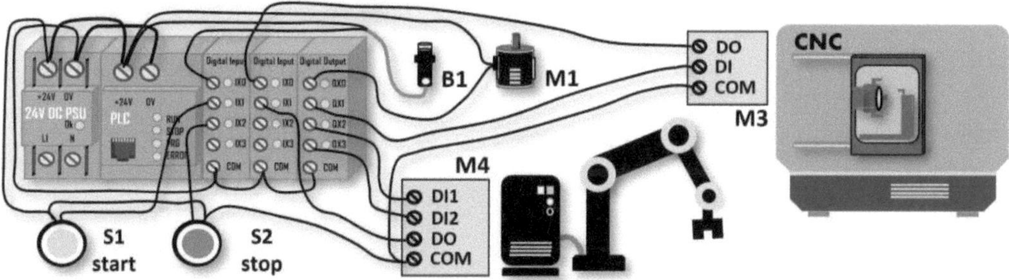


The motor **M1** is used to control conveyor belt **M1**.

When sensor **B1** gives a **TRUE** signal, the conveyor belt must stop so that the robot can take the workpiece from the same position on the conveyor belt.

Conveyor belt **M2** runs continuously.

 

**Below are illustrations of the work process:**

First, the robot picks up a new workpiece from conveyor belt **M1**:

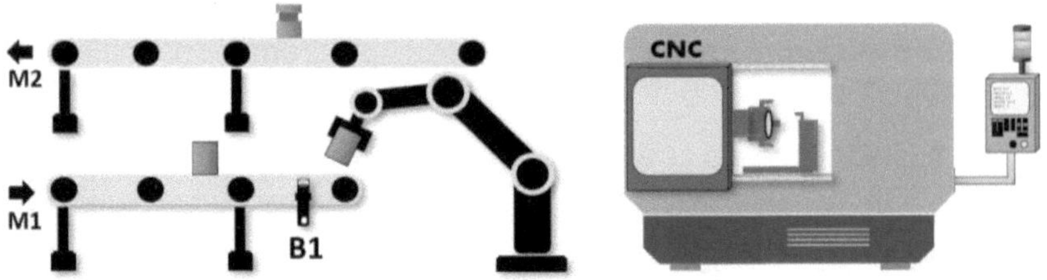

Next, the robot places the workpiece in the CNC machine:

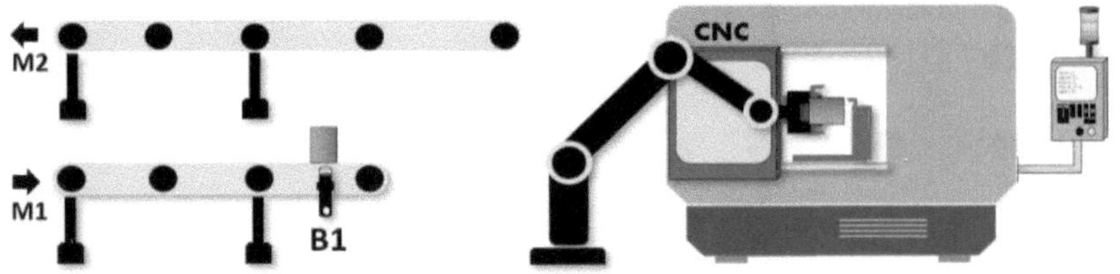

Now the CNC machine processes the workpiece while the robot waits:

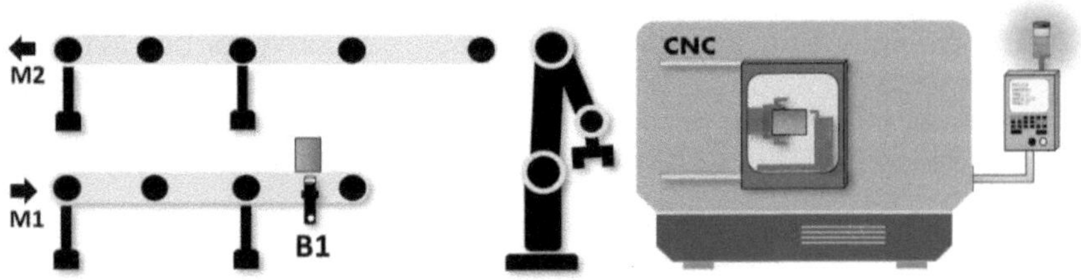

The robot now removes the finished workpiece from the CNC machine:

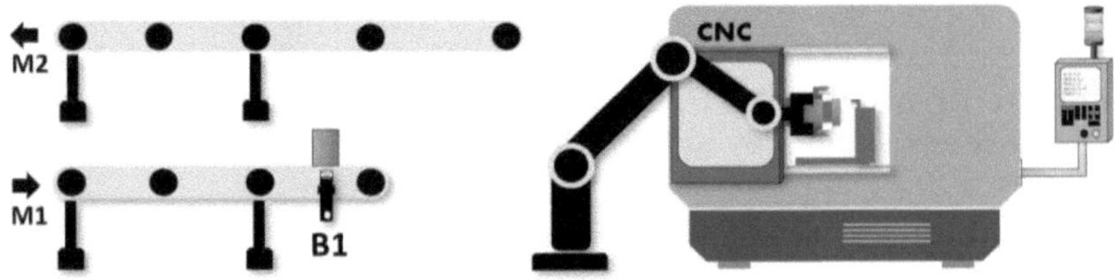

Finally, the robot places the workpiece on conveyor belt **M2**:

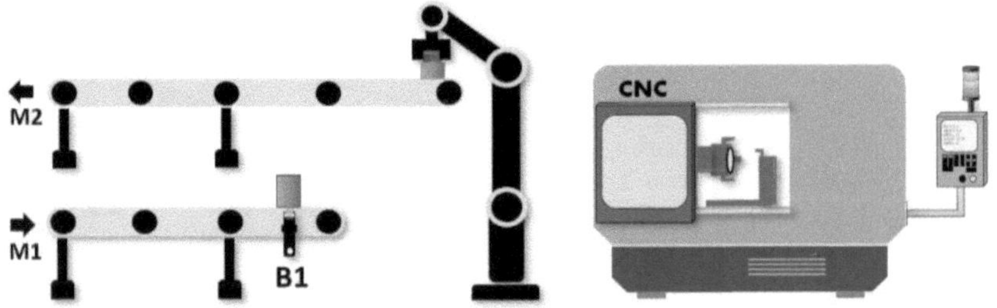

And the entire work process is repeated for the next workpiece.

 

**User operating panel**
The machine is started with the manual push button **S1** and stopped with **S2**.

**S1** is a Normally Open (NO) contact, and **S2** is a Normally Closed (NC) contact.

When the stop button **S2** is pressed, the work continues, and the task is completed (the entire sequence is completed). That is, the system stops only when a processed workpiece is placed on conveyor belt **M2**.

**CNC machine**
The CNC machine is configured (programmed) to be started from an external digital signal **M3_D1**. It starts when the digital signal goes from **FALSE** to **TRUE** (positive edge). When it finishes machining a workpiece, the digital output on the CNC machine **M3_DO** is set from **TRUE** to **FALSE**.

**Robot controller**
The robot controller **M4** controls the robot, and there are two robot programs inside it.

The two programs can be started by external digital signals from the PLC:

| | |
|---|---|
| **Program 1:** | Moving the workpiece from conveyor belt **M1** to the CNC machine. The program starts when the signal on **M4_DI1** goes from **FALSE** to **TRUE**. |
| **Program 2:** | Moving the workpiece from the CNC machine to conveyor belt **M2**. The program starts when the signal on **M4_DI2** goes from **FALSE** to **TRUE**. |

When the robot is busy in moving a workpiece, the signal on the external output of the robot controller **M4_DO** is set to **TRUE**. The robot completes its task when the signal on output **M4_DO** transitions from **TRUE** to **FALSE**.

**Exercise**
Write a PLC program for the small machine setup.

## 2.25 Turn on the light with a two-way switch (toggle switch)

This exercise has a push button switch **S1** and a lamp **P1** connected to a PLC:

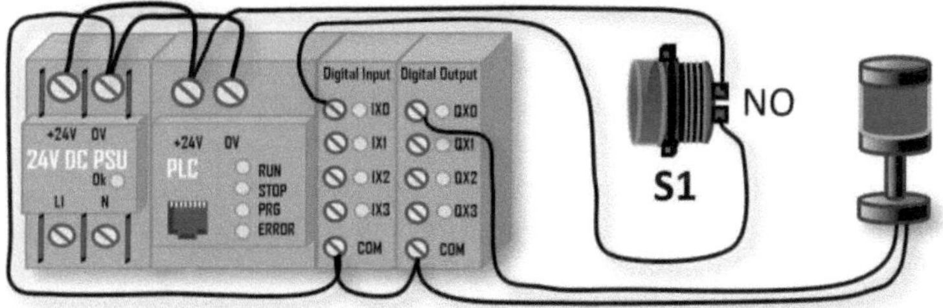


The push button **S1** is a Normally Open (NO) contact with spring return.
Write a program that operates as follows:
When pressing **S1**, the lamp should turn on, and upon another press of **S1**, the lamp should turn off.
Thus, it is possible to turn the lamp on and off using a single push button.

# 3 Programming of safety PLC

This chapter contains exercises that can be programmed in a safety PLC.

## 3.1 Safety light curtain and pallet

In this exercise, you must develop and write PLC code that can turn off and on a safety light curtain, which is also referred to as a light grid.

The setup consists of a conveyor belt used for transporting pallets with packages to an enclosed area where humans do not have access. The area has an opening through which the pallets enter, which is monitored by a light curtain.

The safety light curtain ensures that the opening is monitored by light beams, and if the light beams are interrupted, the machines in the enclosed area must stop.

The light curtain should be turned off when a pallet arrives.

**Images of the process**

This image shows a pallet on its way into the enclosed area.

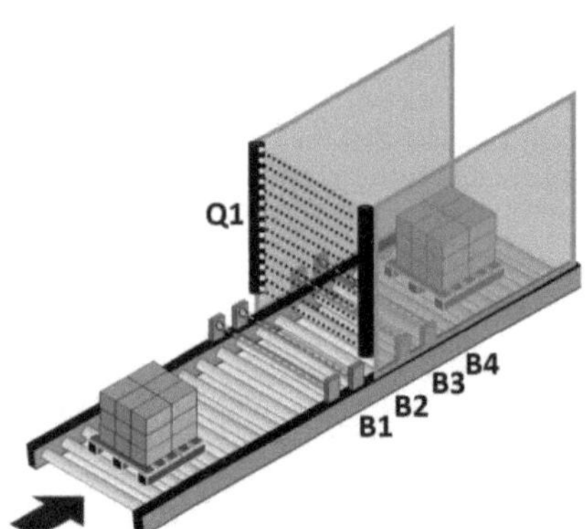


This picture shows the light curtain turned off, and the pallet is moving into the enclosed area.

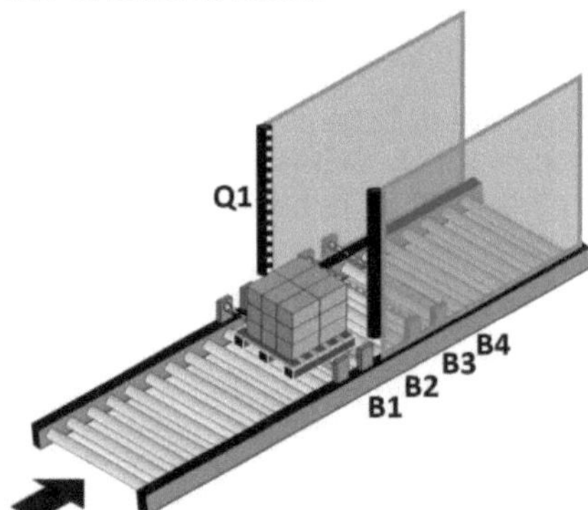


As seen in the images, four sensors are mounted, which can be used to detect the presence of a pallet. The signals from the sensors are used to turn the light curtain off and on according to the following rules:

1) The light curtain must turn off (mute) when both **B1** and **B2** are activated.
2) The light curtain must turn on (unmute) when both **B3** and **B4** are deactivated.
3) The light curtain **Q1** is on when it receives a **TRUE** signal.

The conveyor belt runs constantly in this exercise.

**Exercise**

Write a PLC program to control the light curtain.

 

## 3.2 Robot cell with safety PLC

In this exercise, you will need a robot cell as shown below:

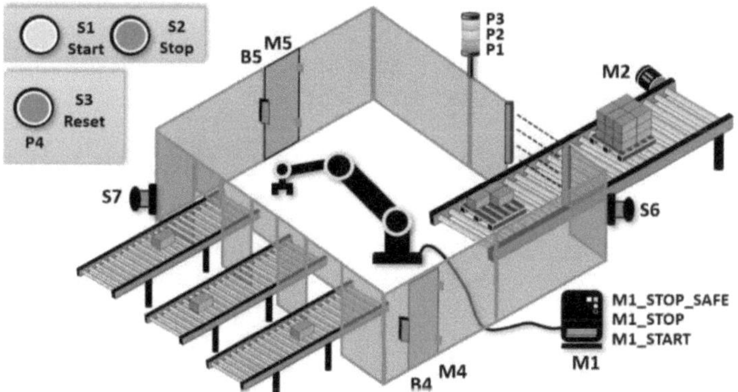


The robot cell has three conveyors that deliver packages to the robot. The robot places the packages on a pallet. When the pallet is full, it is moved out of the robot cell by a conveyor belt **M2**. The robot is controlled by a robot controller **M1**, and it has three digital input signals. All in the robot cell is controlled by a safety PLC.

The following components are included:

| Name | Purpose and operation |
|---|---|
| **M1** | Robot. Should stop when pressing **S2** or emergency stop button. |
| **M1_START** | Input signal. Start robot program.<br>Must not start if emergency stop is activated. |
| **M1_STOP** | Input signal. Stop robot program if **FALSE** signal. |
| **M1_STOP_SAFE** | Input signal. Should be **TRUE** during normal operation.<br>Set to **FALSE** upon pressing emergency stop button. |
| **M2** | Pallet transport motor. Motor is running if **TRUE** signal.<br>Motor should stop when pressing **S2** or emergency stop button. |
| **M4, M5** | Safety door lock. If **FALSE** signal, door in safety enclosure must lock.<br>Door may only be opened if **S2** or emergency stop is pressed. |
| **S1** | Start robot: Set **M1_START** to **TRUE**. Must not start if light on at **P4**. |
| **S2** | Stop robot. Set **M1_STOP** to **FALSE**. |
| **S3** | Reset of emergency stop activation: Light at lamp contact **P4** indicates pressing **S3** is required. |
| **S6, S7** | Emergency stop buttons: If activated, set **M1_STOP_SAFE** to **FALSE**. |
| **B4, B5** | Door contact gives **FALSE** signal when door to safety enclosure is open. |
| **P1** | Light on if emergency stop contact is activated. |
| **P2** | Light on if normal operation. Robot cell is in operation. |
| **P3** | Light on if **S2** is pressed. Turn off when **S1** is pressed. |
| **P4** | Light on if emergency stop is activated. The lamp should turn off when emergency stop is reset. |

**Exercise**

Write a PLC program for a safety PLC to be used for the robot cell.

# 4 Exercises with logic and counters

This chapter contains exercises that can be programmed using logic and counters. The exercises become harder with each step as you go through the chapter; at the end of the chapter there are exercises with encoders.

## 4.1 Turn on lamp after five presses of the push button

In this exercise, a push button **S1** and a lamp **P1** are connected to a PLC:

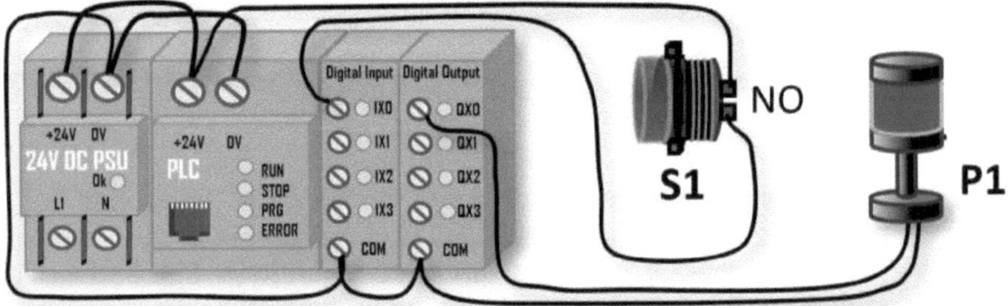


Write a PLC program where lamp **P1** turns on after the push button **S1** has been pressed five times.

## 4.2 Turn the lamp on and off

In this exercise, two push buttons and a lamp are connected to a PLC:

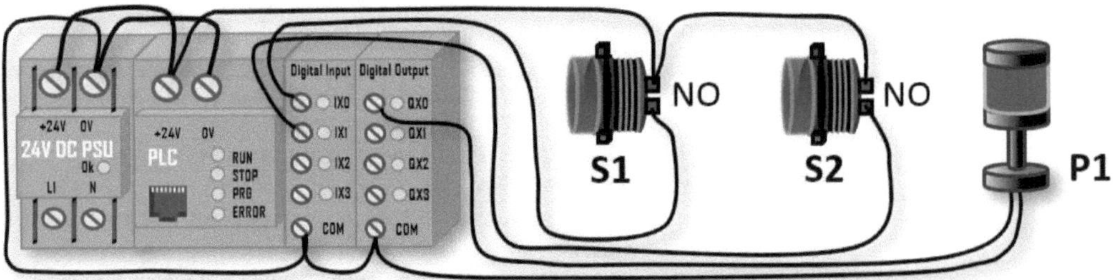


Write a PLC program that meets the following requirements:

- Create a counter that counts the number of times **S1** has been pressed.
- When **S1** has been pressed four times, lamp **P1** should turn on.
- When **S2** is pressed, the counter should reset and lamp **P1** should turn off.

 

## 4.3 Turn the lamp on and off after pressing the push button

In this exercise, a push button **S1** and a lamp **P1** are connected to a PLC:

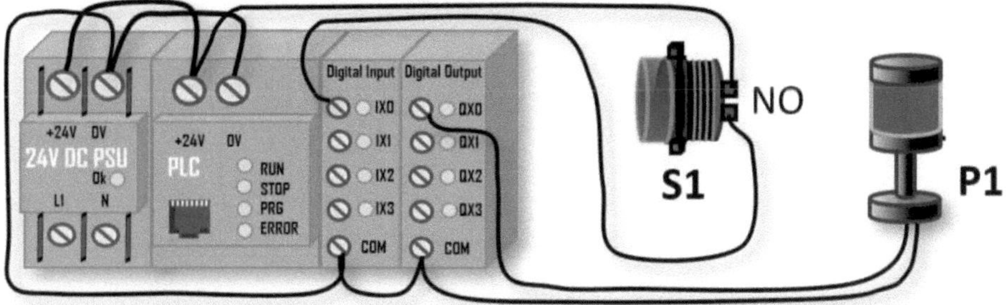


Write a PLC program that operates as follows:

Lamp **P1** turns on after **S1** is pressed five times on and turns off again after an additional five presses on **S1**.
The process repeats: after five presses on **S1**, lamp **P1** turns on, and after five more presses on **S1**, lamp **P1** turns off.

## 4.4 Change lights in the lamp tower

In this exercise, a push button **S1** and a lamp tower are connected to a PLC:

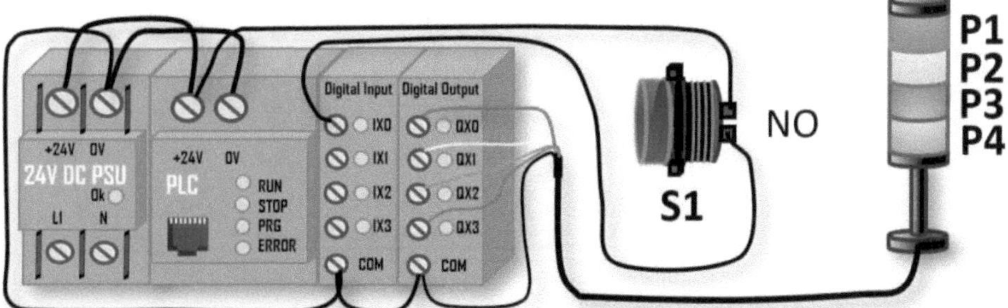


Write a PLC program with the following behaviour:

When the PLC is turned on, there should be no lights on in the lamp tower.

The first time **S1** is pressed, lamp **P1** should turn on. When **S1** is pressed again, the next lamp should turn on. This means both lamps **P1** and **P2** are lit. Pressing **S1** again should turn on lamps **P1**, **P2**, and **P3**. The last time **S1** is pressed, all four lamps should be turned on. Pressing **S1** again should turn off all lamps. The entire sequence should then start over.

## 4.5 Turn on lamp with two push buttons

In this exercise, two push buttons and a lamp are connected to a PLC:

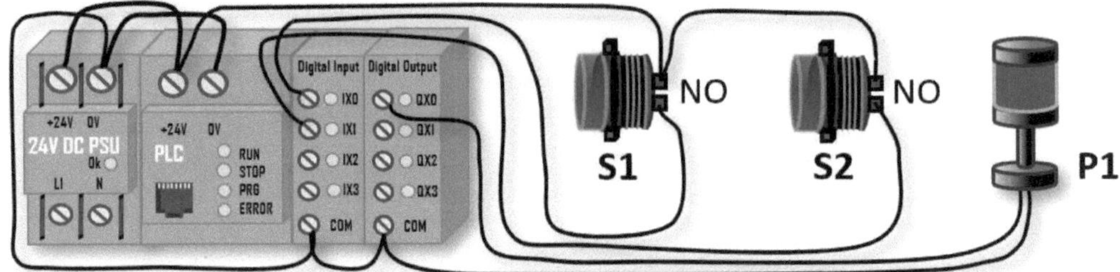


Write a PLC program that meets the following requirements:

- When **S1** is pressed four times followed by five presses on **S2**, **P1** should turn on.
- The lamp must not turn on if either **S1** or **S2** is pressed nine times.
- When both **S1** and **S2** are pressed at the same time, the counter must be reset.

## 4.6 Turn on lamp after multiple presses on the push button

In this exercise, two push buttons and a lamp are connected to a PLC:

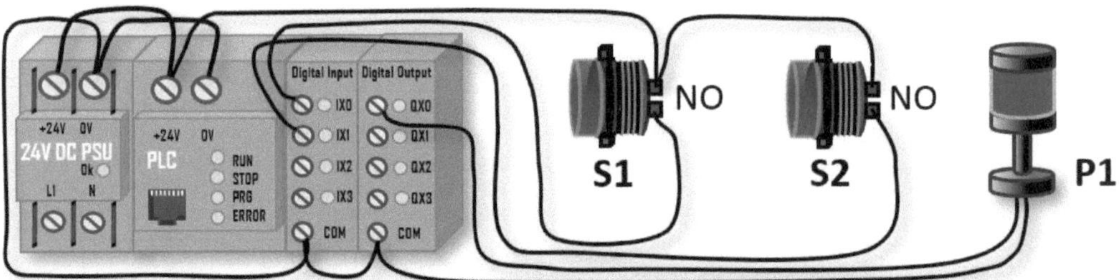


Write a PLC program that meets the following requirements:

- When **S1** is pressed four times followed by three presses on **S2**, **P1** should turn on.
- When both **S1** and **S2** are pressed at the same time, the counter must be reset, and lamp P1 turns off.

The lamp must not turn on if:

- Either **S1** or **S2** is pressed seven times.
- Push button **S2** is pressed three times first followed by four presses on **S1**.

 

## 4.7 PLC program for controlling a parking garage

This exercise develops a PLC program for controlling the access to a parking garage. Below is a drawing of the parking garage area:

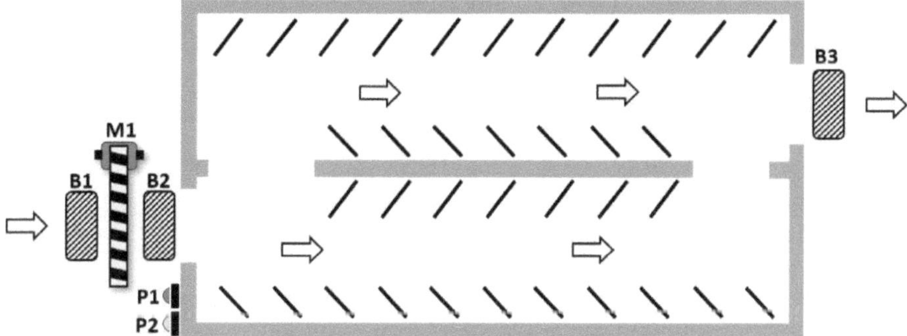


In front of the entrance, there is a barrier, controlled by a motor **M1**. Further, at the entrance, there are two lamps **P1** (red) and **P2** (green), along with two sensors **B1** and **B2**. At the exit, there is a sensor **B3**.

**Car detection**

Sensors **B1**, **B2**, and **B3** are placed under the roadway. Each sensor gives a **TRUE** signal when there is a car directly above it, as shown below:

**The program should operate as follows:**

When there are free spaces in the parking garage, the green lamp **P2** should be turned on, **P1** should be turned off, and the barrier should be closed.
When a car is coming to the parking garage and activates sensor **B1**, the barrier should open; when the car has passed sensor **B2**, the barrier should close, and the number of free spaces should decrease by one.
If **M1** has a **TRUE** signal, the barrier goes up (open), and a car can enter the parking garage. When **M1** receives **FALSE**, the barrier goes down (close).
If the parking garage becomes full, the green lamp **P2** should turn off, the red lamp **P1** should turn on, and the barrier must not be able to go up until one or more cars have left the parking garage.
The number of free spaces increases by one when a car has left the parking garage. A car has left the parking garage when the car at the exit has passed sensor **B3**.

**Exercise**

Write a PLC program for controlling the parking garage.

## 4.8 Grouping of three packages on a conveyor belt

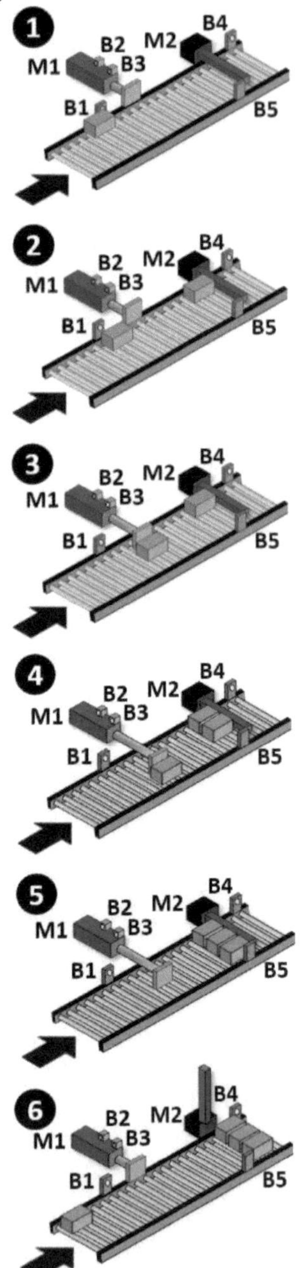


In this exercise, you must write a PLC program that can push packages on a conveyor belt so that three packages are grouped. This allows a robot to pick up three packages at the same time and place them on a pallet.

**Description**

The conveyor belt is equipped with three digital sensors **B1**, **B4**, and **B5**, which give a **TRUE** signal when a package is directly in front of the sensor.

A cylinder **M1** (actuator, an electric cylinder) moves out when **M1** receives a **TRUE** signal. To ensure that the packages are placed correctly on the conveyor belt, the cylinder has two sensors **B2** and **B3**. Sensor **B2** gives a **TRUE** signal when the cylinder has pushed the middle package (package number two) to the correct position in the middle of the conveyor belt. Sensor **B3** gives a **TRUE** signal when the cylinder is moved fully out, where package number three should be.

A mechanical stop **M2** ensures that the three packages are positioned next to each other. It is moved up when **M2** receives a **TRUE** signal. It is moved down again when **M2** receives a **FALSE** signal.

**Explanation of images**

The packages always enter the conveyor belt at the left side. The conveyor belt operates (moves) continuously.

Image (1) shows package number one just activating sensor **B1**. The mechanical stop is down to stop the packages.

Image (2) shows package number one at the mechanical stop, and package number two has just left sensor **B1**. Package number two needs to be pushed onto the middle of the conveyor belt by cylinder **M2**, as seen in image (3).

In image (4), packages number one and two are in front of the mechanical stop, and package number three has just been pushed out to the right side.

Image (5) shows all packages in front of the mechanical stop. Sensor **B5** gives a **TRUE** signal.

Image (6) shows the mechanical stop lifted. It can be lowered again on a falling edge (signal is going from **TRUE** to **FALSE**) from sensor **B4**.

**Exercise**

Write a PLC program based on the description.

 

## 4.9 Chain drive for box lifting

In this exercise, you must develop a PLC program that is able to lift boxes from conveyor belt **M3** to conveyor belt **M4**:

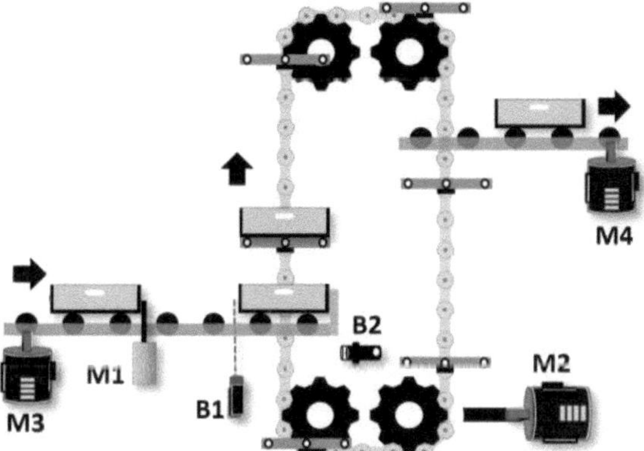


**Description**

The chain drive is powered by a motor **M2**. When **M2** receives a **TRUE** signal, the motor runs, and the chain is pulled around. Six shelves are mounted on the chain, each capable of lifting one box from conveyor belt **M3** to conveyor belt **M4**. The two conveyor belts **M3** and **M4** are running continuously.

The system is called a paternoster elevator.

An actuator (a motor with an arm that can move up) **M1** is used to stop a box while another box is lifted by the chain drive. When **M1** receives a **TRUE** signal, the cylinder moves down, allowing a box to pass by **M1** and move towards the chain drive.
On a positive edge (the digital signal goes from **FALSE** to **TRUE**) from sensor **B1**, cylinder **M1** should move up to stop the next box.

When a shelf on the chain drive is in front of sensor **B2**, the sensor gives a **TRUE** signal. If a box is on its way to where it can be lifted, the chain drive should stop (**M2** is set to **FALSE**) to prevent a collision between a box and a shelf. A box is on its way to be lifted when sensor **B1** has a **TRUE** signal. A box is ready to be lifted after a falling edge (signal goes from **TRUE** to **FALSE**) from sensor **B1**.
When there are no boxes on the chain drive, motor **M2** should stop to save energy. Motor **M2** should only restart when sensor **B1** detects (receives a **TRUE** signal) that a box is ready to be lifted.

**Exercise**

Write a PLC program for the chain drive.

# 4.10 Machine for packing cookies (automatic/manual)

A machine packs cookies into moulded plastic packs, as shown:

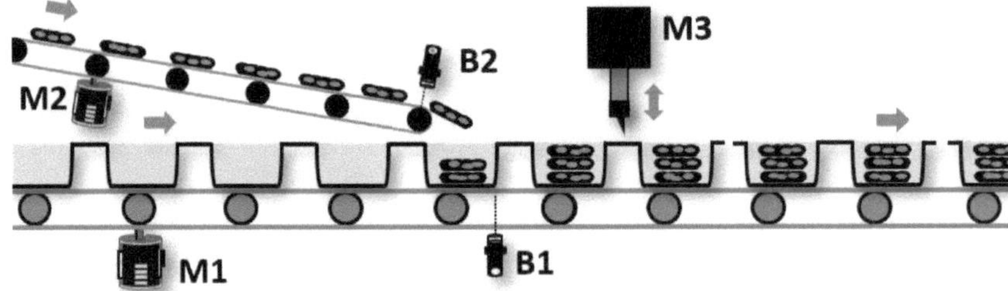


**Description**

Cookies come from the bakery on conveyor belt 2 (the above belt). From the conveyor belt, each cookie falls directly into an endless thermoformed moulded plastic package. When a cookie leaves conveyor belt 2, the digital signal from sensor **B2** changes from **TRUE** to **FALSE**.
Conveyor belt 2 is controlled by motor **M2** and moves to the right with a **TRUE** signal.
The endless plastic package arrives on conveyor belt 1. Sensor **B1** gives a **TRUE** signal when the plastic package is correctly positioned under conveyor belt 2.
Conveyor belt 1 is controlled by motor **M1** and moves to the right with a **TRUE** signal.
There should be three cookies in each plastic package, so conveyor belt 1 should move forward to the right when there are three cookies in the plastic package.
There should be six cookies in each customer package, so the plastic packages are cut with a knife controlled by motor **M3**. When motor **M3** receives a true signal, the knife goes down and cuts the connection between two plastic packages. At a **FALSE** signal, the knife is automatically pulled back with a spring.

**Control panel**

The machine has a control panel containing a rotary switch and five push buttons:

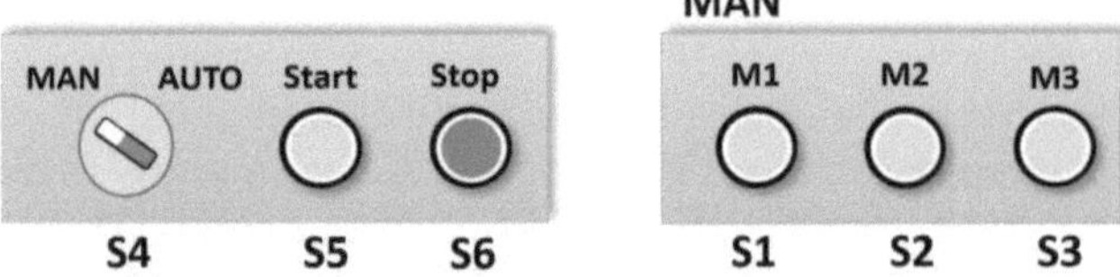


The rotary switch **S4** is used to switch between automatic and manual operation of the machine. In automatic operation, the machine should function as described above.
In automatic mode, the machine can be started with push button **S5** and stopped with push button **S6**.
In manual mode, the motors can be set to **TRUE** using the manual push buttons **S1**, **S2**, or **S3**. If the push button is not held down, the button gives a **FALSE** signal.
Manual operation can be used to test and adjust the machine.

**Exercise**

Write a PLC program for the machine.

 

# 4.11 Filling oil into barrels (encoder)

The setup consists of a small station used for filling barrels with oil:

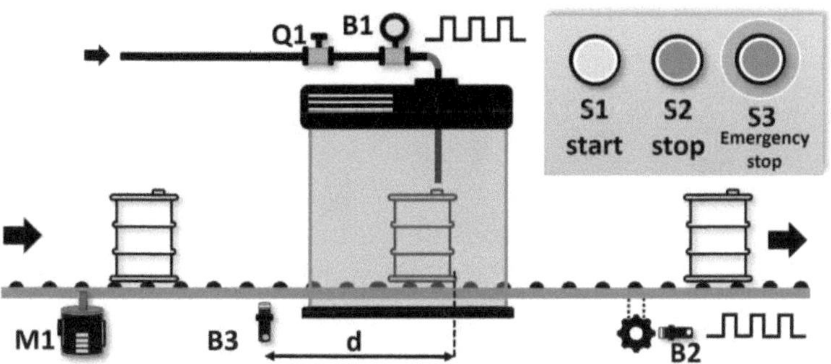


**Description**

A conveyor belt is driven by a motor **M1**. It transports empty barrels into a closed system where the oil filling takes place. The quantity poured into the barrel is measured by a digital flow meter **B1** (this sensor measures the amount of oil passing through the pipe and gives a digital pulse for every 100 ml). Since oil is flammable, there are no electrical components in the closed system.

The barrel must be precisely positioned directly under the filling pipe, so a gear and an encoder sensor **B2** are used, giving digital pulses when the conveyor belt runs.

There can be 800 litres of oil in a barrel.

**Control panel**

The entire station is started with a manual push button **S1** and stopped with **S2**.

When the stop button **S2** is pressed, the barrel is filled, and then conveyed out of the closed system before the conveyor belt stops.

**Component description:**

| Name | Type | Description |
|---|---|---|
| **M1** | Motor | Drives the conveyor belt. Set to **TRUE** to run the conveyor. |
| **Q1** | Valve | Electric open/close valve for oil filling. Set **TRUE** to open. |
| **B1** | Sensor | Measure oil quantity. One pulse for every 100 millilitres [ml]. |
| **B2** | Sensor | Encoder sensor. Roller diameter: 15 cm, Distance d = 120 cm. |
| **B3** | Sensor | Gives a **TRUE** signal when a barrel is at the sensor. |
| **S1** | Push button | Start filling station. Normally Open (NO) contact. |
| **S2** | Push button | Stop filling station. Normally Closed (NC) contact.<br>Barrel filling completes and barrel is moved out. |
| **S3** | Push button | Stop system immediately. Valve **Q1** closes and motor **M1** is stopped (set to **FALSE**). If **S2** is pressed afterwards, the barrel is moved out. |

**Exercise**

Write a PLC program based on the description.

# 4.12 Machine for cutting cables (encoder)

In this exercise, you must develop a PLC program for a machine that cuts cables to a specific length. Image of the machine:

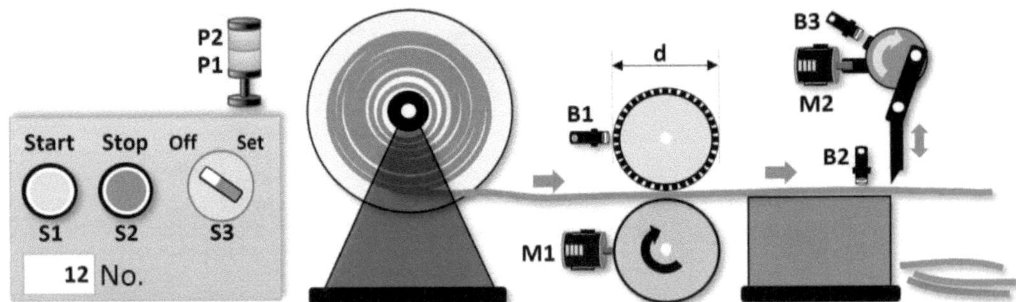


**Description**

The cable is pulled forward by a motor **M1**, which runs at a constant low speed when there is a **TRUE** signal on **M1**.

A motor **M2** is used to cut the cable. With a **TRUE** signal on **M2**, the motor should rotate once turn to cut the cable. Sensor **B3** gives a **TRUE** signal when the knife is up and the cable can be pulled forward.

Sensor **B2** gives a **TRUE** signal when there is a cable present.

An encoder sensor **B1** is used to measure the length of the cable.

The diameter **d** is 15 cm.

**Light tower**

A light tower with two lamps **P1** and **P2** is used to inform the operator working at the machine. When light is turn on in **P2**, the machine has finished cutting cables. When light is turned on in **P1**, the machine is in automatic operation and cutting cables.

**Control panel**

The control panel contains a start button **S1**, which starts the machine and cuts the number of cables specified in the "No." input field.

The stop button **S2** immediately stops **M1** and **M2**. If the machine has not finished cutting the number of cables specified in the "No" field, it continues to cut cables if the start button **S1** is pressed after a stop.

If the machine is not running (light in **P1**), the cable length can be adjusted with the rotary switch **S3**. This is done by setting **S3** to the "Set" position, and then pulling the cable forward to the desired length. When **S3** is set back to the "Off" position, the measured length is saved in the PLC, becoming the new cable length.

**Exercise**

Write a PLC program based on the description.

 

# 5 Exercises with timers

This chapter contains PLC program exercises where you learn to use timers. Some of the exercises also require the use of logic and counters in the PLC program.
The exercises get harder as you progress through the chapter.

## 5.1 A rotary switch turns on a lamp after 10 seconds

In this exercise, a rotary switch **S1** and a lamp **P1** are connected to a PLC:

Write a PLC program where lamp **P1** turns on 10 seconds after rotary switch **S1** is set to the "1" position. When the rotary switch is set to "0", the lamp **P1** should turn off.

## 5.2 A lamp should be turned on for 10 seconds

In this exercise, a push button **S1** and a lamp **P1** are connected to a PLC:

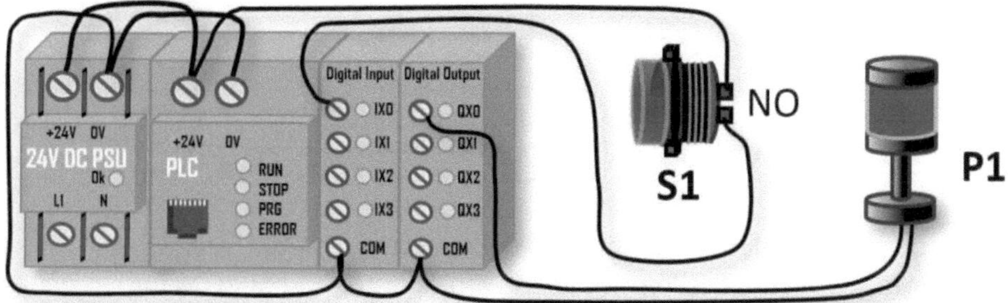


Write a PLC program where lamp **P1** turns on immediately when button **S1** is pressed. The lamp should remain on for 10 seconds and then turn off.

The lamp **P1** should only be turned on for 10 seconds, even if push button **S1** is held down longer than 10 seconds.

## 5.3 Use a rotary switch to make a lamp flash

Here is an exercise with a manual rotary switch **S1** and a lamp **P1**:

Write a PLC program with the following behaviour:

When rotary switch **S1** is in position "1" (on), the lamp **P1** should flash. This means the lamp should be on for 2 seconds and off for 2 seconds.

## 5.4 Use a rotary switch to make two lamp flash

Here is an exercise with a manual rotary switch **S1** and two lamps:

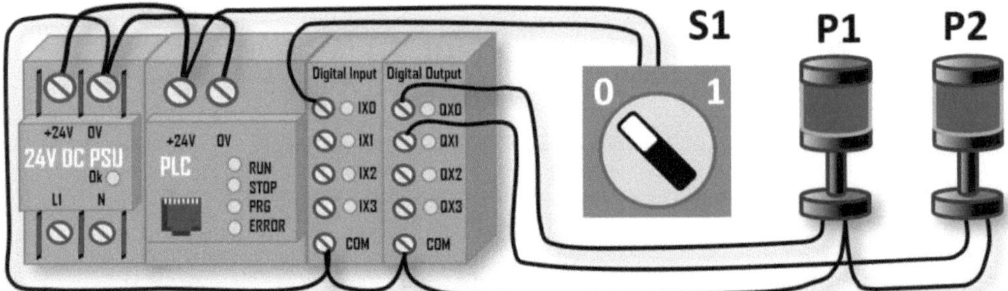


Write a PLC program that meets the following requirements:

- When rotary switch **S1** is in position "1" (on), lamp **P1** should be turned on.
- After 10 seconds, lamps **P1** and **P2** should both be turned on.
- After another 10 seconds, the two lamps should alternate flashing with a one-second interval. This means that when **P1** is off, **P2** is on, and vice versa.

 

## 5.5 Push buttons that change the speed of lamp flashing

In this exercise, two push buttons and a lamp are connected to a PLC:

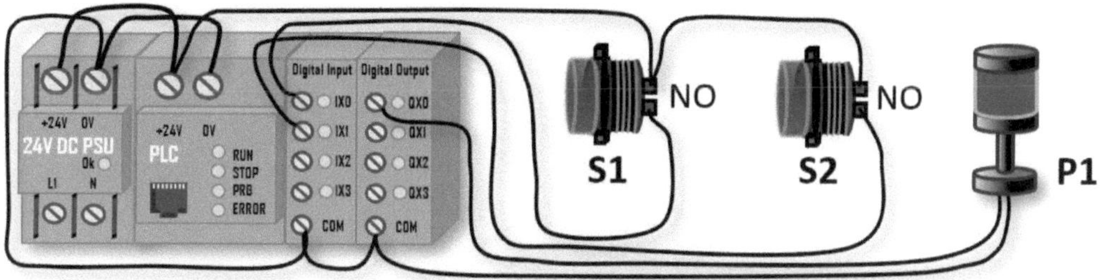


Write a PLC program that meets the following requirements:

- When pressing **S1**, the lamp **P1** flashes faster and faster until it stays constantly on.
- When pressing **S2**, the lamp **P1** flashes slower and slower until it's completely off.

## 5.6 Light tower used as a running light

In this exercise, a push button **S1** and a light tower are connected to a PLC:

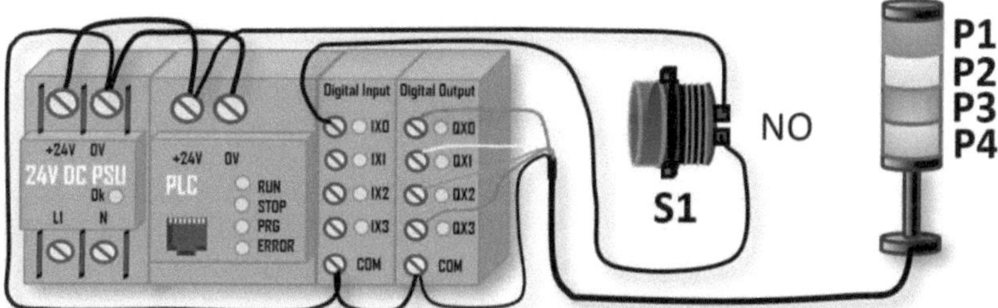


Write a PLC program that meets the following requirements:

When button **S1** is pressed, the lights should alternate (rotate through) as follows in the lamps: First, **P1** turns on. **P1** turns off and **P2** turns on. **P2** turns off and **P3** turns on. **P3** turns off and **P4** turns on. **P4** turns off and **P1** turns on.
And the cycle repeats as it is a running light.
A lamp should be on for 2 seconds.

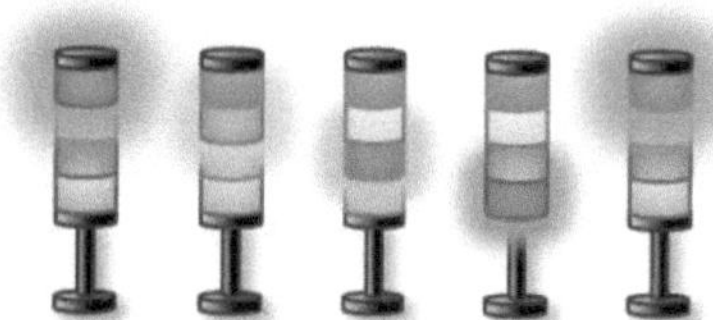

 

## 5.7 Pedestrian crossing traffic signal

In this exercise, you must write the PLC program for controlling a traffic signal at a pedestrian (person) crossing field, which looks like this:

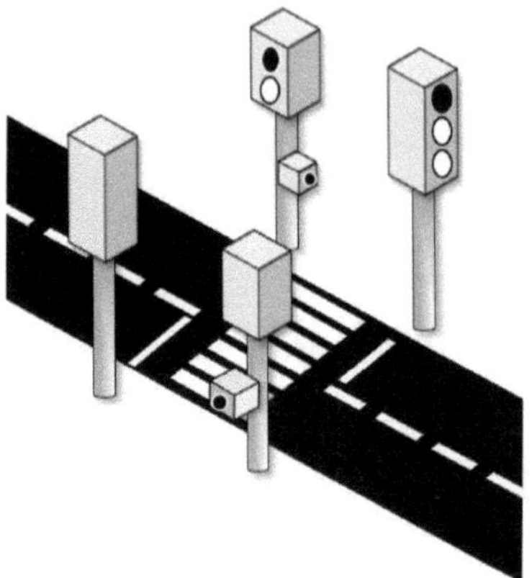

**Description**

There is a constant green light for vehicles. When a person wants to cross the road, they must press the push button on the pole on either side of the road. The two push buttons are named **S1** and **S2**. The person may only cross the road when there is a red light for vehicles and a green light for persons.

The sequence for the traffic signal can be divided into these six steps (states) as shown below:

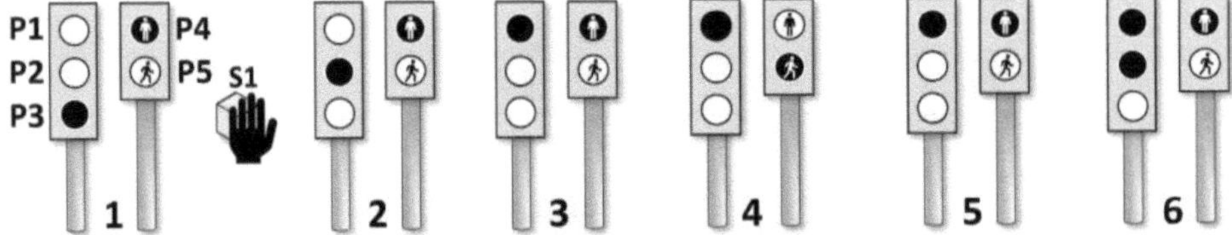


Table showing the times between each step:

| Transition from step | Transition to step | Change condition |
|---|---|---|
| 1 | 2 | Upon pressing **S1** or **S2** |
| 2 | 3 | 8 seconds |
| 3 | 4 | 10 seconds |
| 4 | 5 | 30 seconds |
| 5 | 6 | 5 seconds |
| 6 | 1 | 3 seconds |

**Exercise**

Write a PLC program for the traffic signal.

 

# 5.8 PLC program for controlling a drawbridge

In this exercise, a PLC program system needs to be developed for a drawbridge:

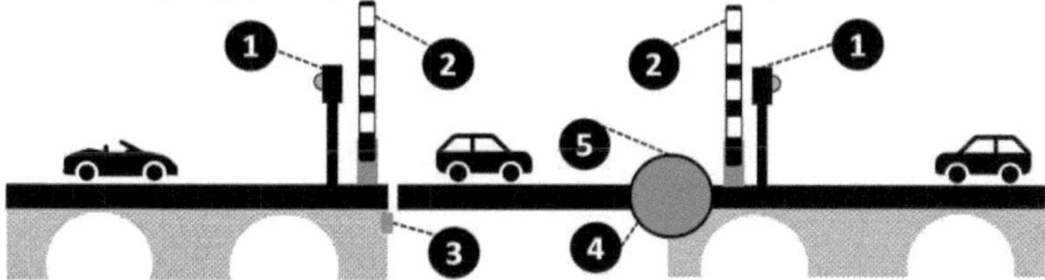


**Description**

The two signal lamps (1) are controlled by separate digital outputs from the PLC. If the signal on the digital output is **TRUE**, the signal lamp is turned on.
The two barriers (2) are controlled by separate digital outputs from the PLC. The barriers will swing down and stop the traffic when there is a **TRUE** signal on the digital output. When the digital output has a **FALSE** signal, the barriers will swing back, and the traffic can pass again.
There is a sensor (3) that gives a **TRUE** signal when the drawbridge is down, and another sensor (5) gives a **TRUE** signal when the drawbridge is up.
The drawbridge itself is controlled by a large motor (4) for raising and lowering it. The motor is controlled by two digital signals from the PLC. One signal is used to start the motor, and the other is used to set the direction of the motor, so the drawbridge either goes up or down. The bridge goes up when both signals are **TRUE**. The motor must stop moving if push button **C)** is activated or if there is a **TRUE** signal from sensor (3) or a **TRUE** signal from sensor (5).
Here is the bridge open:

**Control Panel**

To operate the drawbridge control, there need to be three manual push buttons:

**A)** Open drawbridge, **B)** Close drawbridge, **C)** Stop drawbridge movement.

When manual push button **A)** is activated, signal lamps should flash, and after 30 seconds, the barriers should close the bridge. Then the drawbridge opens. Signal lamps should flash with 1 second on and 1 second off.
When manual push button **B)** is activated, the drawbridge should lower again. When the drawbridge is down, the barriers should open again. The barriers should take 5 seconds to move. When the bridge is open for traffic again, lamps should turn off.

**Exercise**

Write a PLC program for the drawbridge control.

## 5.9 Bag on conveyor belt

In this exercise, you must write a PLC program for a small conveyor where a bag is moved forward on a conveyor belt to be filled and then moved back again.
There is a conveyor belt driven by a motor **M1** and two sensors **B1** and **B2**. The two sensors are mounted at each end of the conveyor belt. Further, there is an operator panel with a start button **S1**, a stop button **S2**, and a button **S3**, which pauses the operation. Below is an illustration:

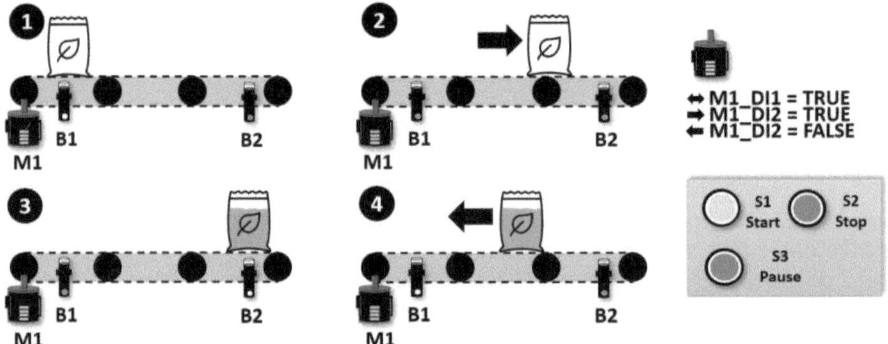


**Description**

The sensors **B1** and **B2** give a **TRUE** signal when the bag is directly above the sensor. When stating the bag is placed on the conveyor belt at sensor **B1**. At sensor **B2**, the bag is filled. Motor **M1** is controlled by two digital input signals: **M1_DI1** and **M1_DI2**. When **M1_DI** is **TRUE**, the motor is turned on and the conveyor belt moves. It can change direction with the digital signal **M1_DI2**. If **M1_DI2** is **TRUE**, the bag moves to the right, and if the signal is **FALSE**, the bag moves to the left.

**Operation**

When the bag is placed at sensor **B1** and the start button **S1** is pressed, the program starts (see image ❶).
Once the program is started, the bag moves to the right (see image ❷).
When the bag reaches sensor **B2**, it needs to be filled, which takes 10 seconds (see image ❸). Afterward, the motor should change direction so that the conveyor belt moves the bag back (see image ❹).
When the bag returns to the starting point ❶, the conveyor belt should stop.
If the operator presses the stop button **S2**, the operation must stop, and the bag must be removed manually from the conveyor belt. The operation can only be restarted by placing a new bag at sensor **B1** and pressing the start button **S1**.
If the operator presses the pause button **S3**, the operation stops immediately and can be restarted at **S1** without the need for a bag at sensor **B1**. A bag already on the conveyor belt continues from where it left off.

**Exercise**

Write a PLC program based on the description.

 

# 5.10 Painting machine

In this exercise, you must write a PLC program for a machine that paints workpieces:

**Description**

The conveyor belt is controlled by a motor **M1**, which can move a workpiece to the right by setting **TRUE** on **M1_K1** and to the left by setting **TRUE** on **M1_K2**. The signals to motor **M1_K1** and **M1_K2** cannot be set to **TRUE** at the same time.
A workpiece to be painted is placed at sensor **B1**. When there is a workpiece at sensor **B1**, it gives a **TRUE** signal, and the machine can be started by pressing **S1**. The workpiece is then moved into the painting machine. When sensor **B2** gives a **TRUE** signal, the workpiece is inside the painting machine. Now, motor **M4** is started for ventilation, and the painting process begins by setting the spray gun **M2** to **TRUE**. Painting continues for 20 seconds, then the workpiece needs to dry, which is done by setting **M3** to **TRUE**. Drying takes 30 seconds. Finally, the workpiece waits for 10 seconds in the machine while ventilation **M4** is still operating. Then ventilation **M4** is turned off, and the painted workpiece is moved back to sensor **B1**.
If the operator presses the stop button **S2** while a workpiece is being painted or dried, the workpiece should be disposed of. This is done by setting **M1_K1** to **TRUE** until sensor **B3** gives a **TRUE** signal. The workpiece is then placed in the pallet for rejected workpieces.

**Lamp tower and machine operation**

A lamp tower with three lamps **P1**, **P2**, and **P3** is used to provide information to the operator working at the painting machine. When the light is turned on in **P1**, the painting machine is ready to receive a new workpiece.
When the operator presses **S1**, the machine goes into operation; lamp **P2** must be turned on and **P1** must turn off.
If the operator presses **S2** while the machine is processing a workpiece, **P3** should be turned on while the workpiece is moved out to the pallet of rejected workpieces.

**Exercise**

Write a PLC program based on the description.

## 5.11 Motor control with relays and tachometer hour counter

In this exercise, you must write a PLC program to control a motor.
The setup is as shown below:

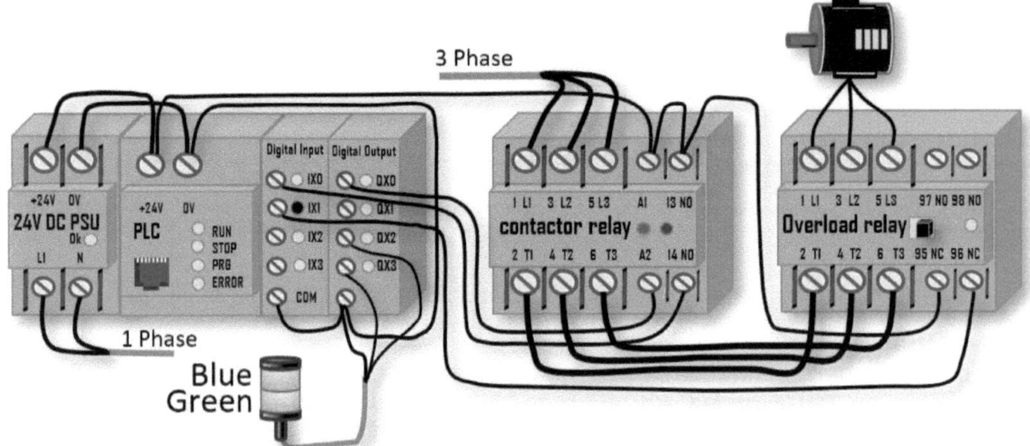


**Components in the setup**

There is a contactor relay, overload relay (overload protection relay), a motor, and a light tower with two lamps. The components are all connected with wires and cables as shown in the above drawing. There is a digital feedback signal to the PLC from both the overload protection relay and the contactor relay.

Further, there is a manual push button switch:

**Requirement specification**

When the Power button is pressed, the motor should run (the motor is in operation), and when it is pressed again, the motor should stop. The Power button is a normally open contact with spring return, but it should function as a toggle switch in the PLC program.

When the motor is running, the green lamp should be on.
When the motor is stopped, the blue lamp should be on.

The following functions should be included in the PLC program:

- A counter that counts the number of times the motor has been started.
- A tacho hour counter (tachometer), which counts the number of seconds and minutes the motor has been in operation.

**Exercise**

Write a PLC program based on the description.

 

## 5.12 Time control of mixer in a process tank

In this exercise, you must write a program for time control of a mixer in a process tank. The setup is as shown below:

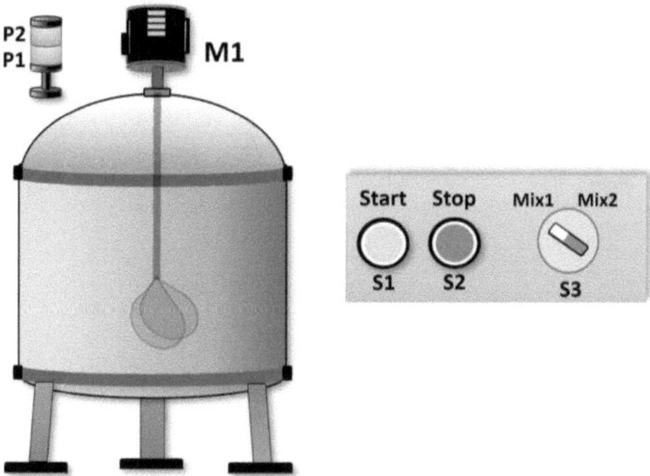


**Description**

At the top of the process tank is a motor **M1**, which drives a mixer.
The automatic operation period starts with the manual push button switch **S1** and stops with the manual push button switch **S2**.
**S1** is a Normally Open (NO) contact, and **S2** is a Normally Closed (NC) contact.

**Automatic operation**

The operation period is 5 minutes, during which there should be light in lamp **P1**.
When the 5 minutes have elapsed, **P1** should turn off, and **P2** should turn on.
During the operation period, the mixing process can be set to one of two modes:

**Mix1**: Mixer is on for 10 seconds and off for 20 seconds.
**Mix2**: Mixer is on for 20 seconds and off for 10 seconds.

An operator can choose between the two modes using manual rotary switch **S3**.

Pressing the stop button **S2** pauses the timer, and operation resumes the process with a press on the start button **S1**.

**Exercise**

Write a program based on the description.

## 5.13 Pallet magazine

In this exercise, you must develop a PLC control for a pallet magazine; which contains a stack of pallets and is used in many companies with production operations. It is a small machine that can automatically supply a new pallet from a stack of them.

Here is an illustration of the pallet magazine:

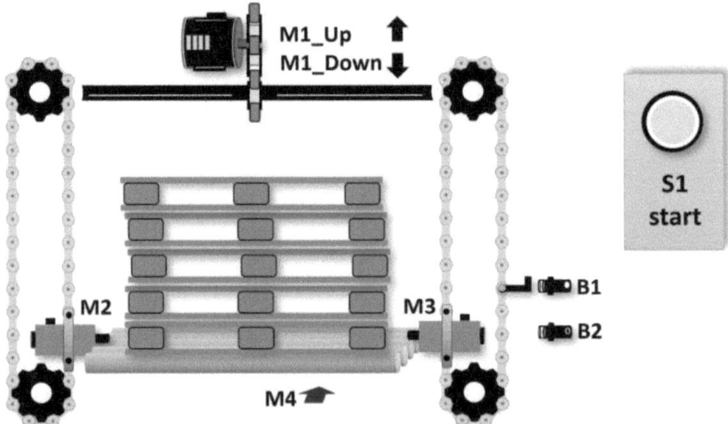


**Description**

Every time a pallet is needed, an operator must press the manual switch **S1**.
It is always the bottom pallet that is taken from the stack of pallets.
The pallet magazine has a motor controlled **M1** chain drive, which can lift the entire stack of pallets.

**Sensors**

**B1** gives a **TRUE** signal when the stack of pallets must not be moved further down.
**B2** gives a **TRUE** signal when the stack of pallets must not be moved further up.
There is no sensor that gives a signal when there are no more pallets in the stack.

**Cylinders**

Two cylinders **M2** and **M3** are used to hold the entire stack of pallets while the bottom pallet is moved away on conveyor belt **M4**, which ensures that the pallet is taken to where it is needed. Conveyor belt **M4** may only operate (run) when the pallet needs to be moved away.

The two cylinders **M2** and **M3** move towards the stack of pallets when the cylinders receive a **TRUE** signal. They must receive a **TRUE** signal for 10 seconds before it is safe to assume that the cylinders are fully pressed against a pallet.

 

Below are six images of the pallet magazine in various parts of the process:

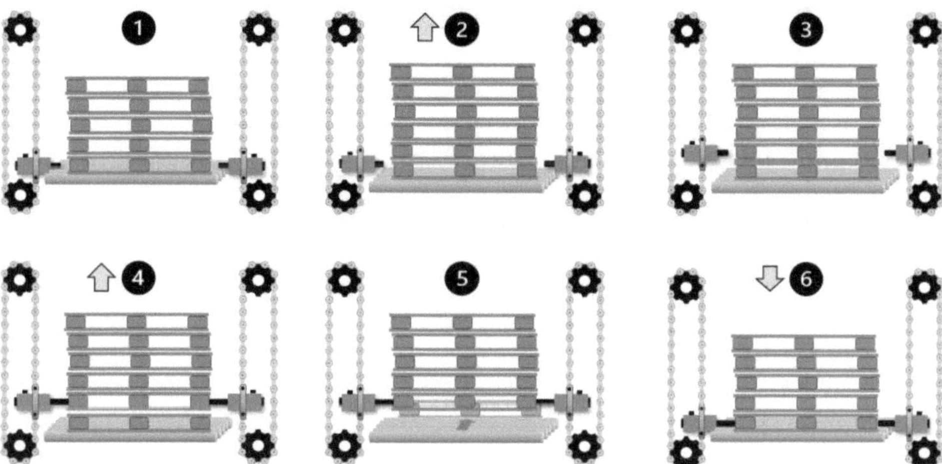


Explanation for the images:

❶ Starting. The entire stack of pallets is on the conveyor belt.

❷ An operator has pressed **S1** to get a pallet. The two cylinders are now to be moved up to the second-bottom pallet. This is done by setting **M1_Up** to **TRUE**. The upward movement stops when sensor **B2** receives a **TRUE** signal because it means the cylinders are positioned in front of the second bottom pallet in the stack.

❸ The two cylinders are now in front of the second bottom pallet, and can be moved in towards the pallet. This is done by setting the cylinders **M2** and **M3** to **TRUE**. The pallet stack may be lifted after 10 seconds, as it takes time for the cylinders to be fully engaged with the pallet.

❹ The pallet stack is lifted, except for the bottom pallet. The pallet stack only needs to be lifted a small distance, which is done by setting **M1_Up** to **TRUE** for 5 seconds.

❺ When the pallet stack is lifted, the bottom pallet can be moved away on conveyor belt **M4**. It takes 12 seconds to move the pallet away. The pallet must be removed before the entire stack of pallets can be lowered onto the conveyor belt again.

❻ The pallet stack is lowered again by setting **M1_Down** to **TRUE**. The cylinders should not come all the way down to conveyor belt **M4**, as this could damage the pallet magazine. Therefore, **M4** should stop when sensor **B1** gives a **TRUE** signal. Then, the two cylinders are set to a **FALSE** signal so that the cylinders no longer hold onto the pallet stack. It takes 10 seconds for the cylinders to move away from the pallet stack.

**Exercise**
Write a PLC program for the pallet magazine.

# 6 Exercise with analog signals

This chapter contains exercises where an analog input value from, for example, a sensor needs to be converted using linear scaling. The value from the analog input module must be converted (scaled) so that the value is usable in a PLC program. In all exercises, a 12-bit analog input module is used.

## 6.1 Analog signal from a temperature sensor

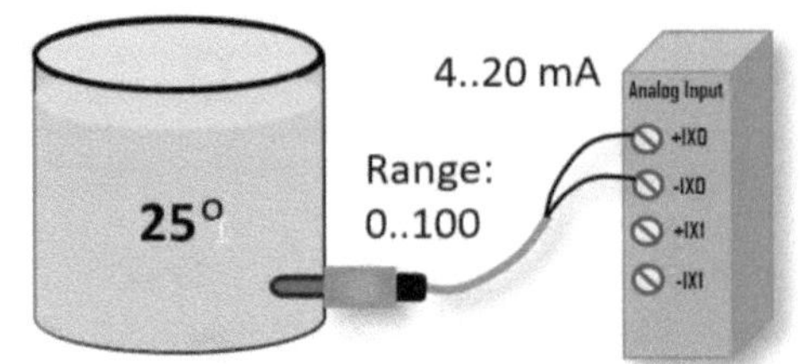


A temperature sensor has a measurement range of 0 to 100 degrees and is connected to a 4-to-20 mA analog input.

The temperature sensor measures 25 degrees in the tank.

What value (number) can be read from the analog input module?

Write a PLC code that can scale the value from the analog input module to 25.

## 6.2 Analog signal from a potentiometer

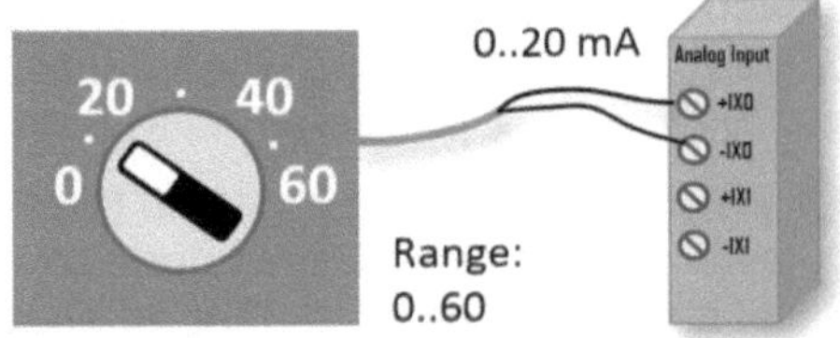


A potentiometer on a user panel can be set to a value in the range of 0 to 60 and is connected to a 0-to-20 mA analog input.
An operator has set the potentiometer to 10.
What value (number) can be read from the analog input module?
Write a PLC code that can scale the value from the analog input module to 10.

## 6.3 Analog signal from a weighing scale

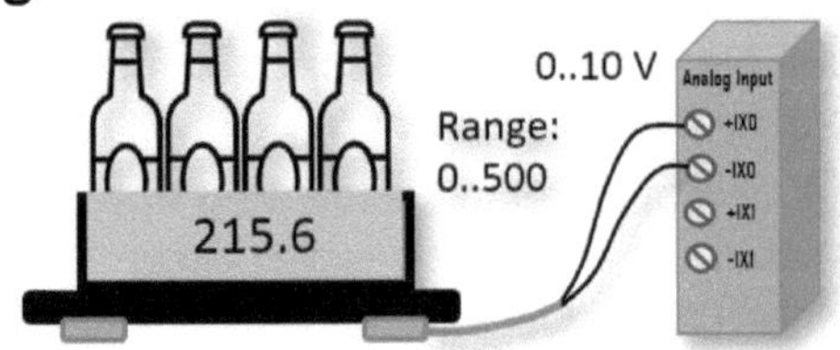


A box of empty bottles is placed on a weighing scale. An analog output from the weighing scale is connected to a 0-to-10 V analog input.
The box with bottles weighs 215.6 g.
What value (number) can be read from the analog input module?

Write a PLC code that can scale the value from the analog input module to 215.6.

 

Below are exercises in which an analog output value needs to be converted using linear scaling. This is relevant when a PLC program needs to control external devices such as the speed of a motor.
In all exercises, a 12-bit analog output module is used.

## 6.4 Value for frequency converter (VFD)

The speed of a frequency converter is controlled from a PLC. An analog output from the PLC is connected to an analog input on the frequency converter.
Control range: 0 to 50 Hz.

The frequency converter should run at 47 Hz. What value should the PLC program write to the analog output?

Write the program that can convert 47 to the value that needs to be written to the output.

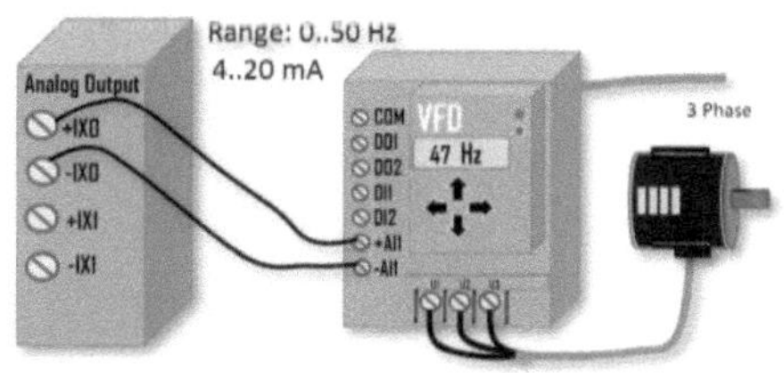


## 6.5 Value for powder dosing

A motor for powder dosing is controlled from a PLC. An analog output from the PLC is connected to an analog input on the motor. Control range: 0 to 120 RPM (revolutions per minute). The motor should run at 30 RPM. What value should the PLC program write to the analog output?

Write a program that can convert 30 to the value that needs to be written to the output.

## 6.6 Value for control valve

A control valve (also referred to as a positioner or regulating valve) for regulating liquid flow is controlled from a PLC.
An analog output from the PLC is connected to an analog input on the control valve.
Control range: 0 to 100% (0% = closed).
The valve needs to be open at 15.5%.

What value should the PLC program write to the analog output?

Write a PLC program that can convert 15.5 to the value that needs to be written to the output.

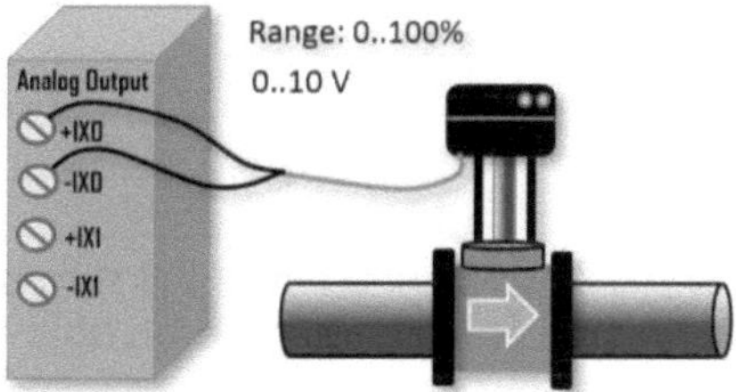

## 6.7 Weighing and counting of suitcases at an airport

In this exercise, you must develop and write a PLC program that can weigh and count the number suitcases at an airport. The setup is as shown below:

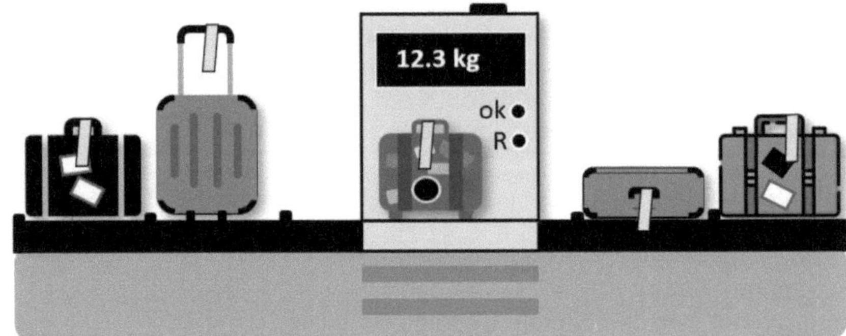


**Description**
Each suitcase is weighed on a scale.
The scale used for a suitcase in the range 0 to 50 kg.
The scale is connected to a PLC via an analog input.
On the front of the scale, is a user panel with two manual push buttons named "**ok**" and "**R**". Both buttons are Normally Open (NO) contacts.
When the suitcase is placed on the weighing scale plate, the airport operator must press "**ok**" to register the measured value.

**Requirements for the PLC program:**

1) It must include a counter, showing the total number (quantity) of suitcases that have been weighed.
2) The program should calculate the average weight of all suitcases.
3) A suitcase belongs to a weight class according to the following table:

| Suitcase size | Weight class | Quantity |
|---|---|---|
| 0 to 5 kg | Small size | ? |
| 5 to 10 kg | Normal size | ? |
| 10 to 20 kg | Big size | ? |
| Bigger than 20 kg | Over size | ? |

The program should count the number of suitcases (quantity) in each weight class.
The airport operator should be able to reset all counters by pressing the "**R**" button, which means reset (set to zero).

**Exercise**
Write a program based on the description and requirements.

 

## 6.8 Luxury hand dryer with time control and progress bar

In this exercise, you must develop a PLC program for a luxury hand dryer. It is found in many restrooms and offers the advantage of eliminating the need for paper towels or hand towels to dry hands.

It is a luxury hand dryer with a heating element **E1** and the option to set the duration time for which the hand dryer should remain turned on:

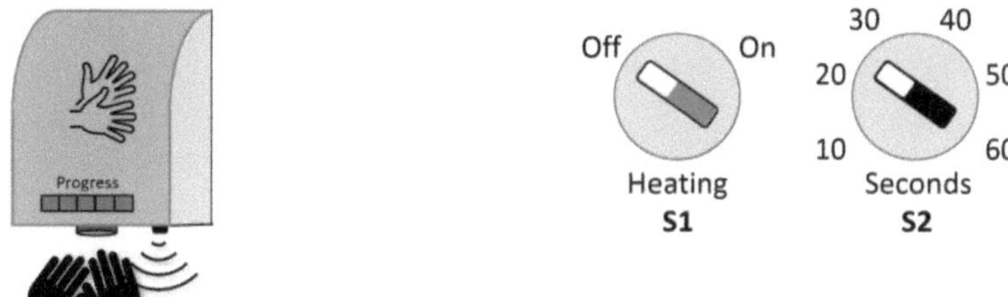


**Description**

The hand dryer has an integrated air blower driven by a motor **M1**. Air comes out from the bottom of the dryer, where hands can be dried.

At the bottom of the dryer there is a sensor **B1** that detects hand movements. When the sensor detects hands, the sensor gives a **TRUE** signal, which is used to start the motor. The motor should run for the duration time set on the analog potentiometer **S2** (between 10 and 60 seconds). The motor should automatically turn off when the set time has elapsed.

**Components inside the hand dryer**

The hand dryer contains the mentioned components and a small PLC:

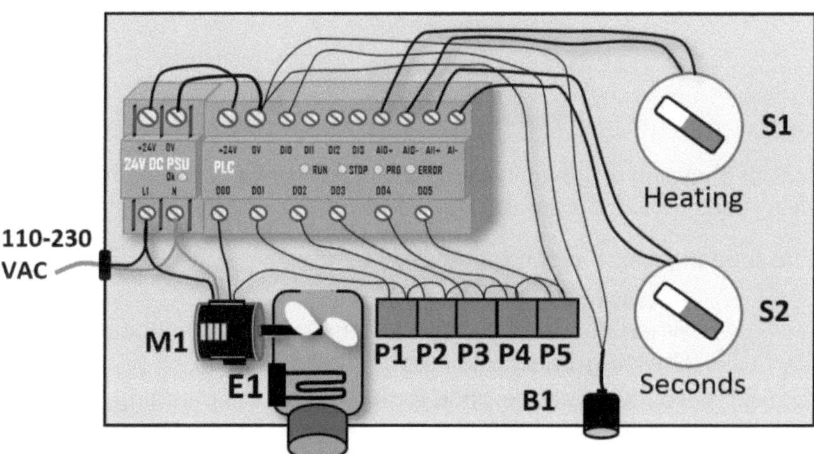


 

**Progress bar**

The hand dryer includes a progress bar with five lamps (**P1** to **P5**), allowing the user to see how much time is remaining. When the dryer is activated (see image 1), all five lamps are turned on, and as time passes, the lamps turn off one by one. When all lamps are turned off (see image 3), the hand dryer stops:

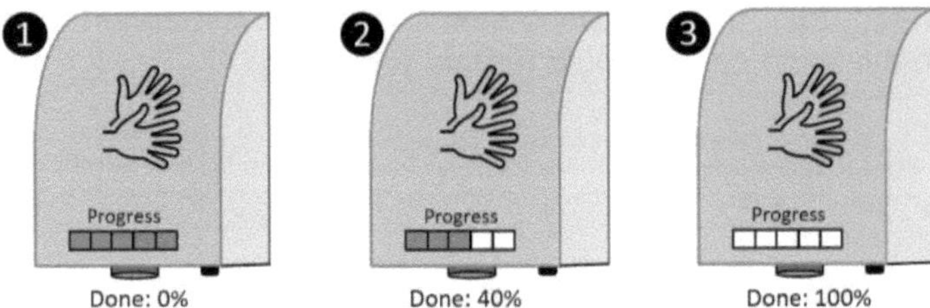


**Heating element**

The heating element **E1** may only be turned on together with motor **M1**, and the heat can be deselected by setting the switch **S1** in the "Off" position.

**Exercise**

Write a PLC program for the hand dryer.

## 6.9 Calculation of electric car charging cost

In this exercise, you must write a PLC program that can calculate the cost of charging an electric car:

To calculate the cost of charging an electric car, current and voltage must be measured. Current is measured in amperes [A], and voltage is measured in volts [V].

To start and stop the measurement, a control panel is used:

**Control panel**

When the user presses **S1**, the calculation should start. Pressing **S2** should stop the calculation.

To calculate the price to be paid for the charging, the amount used [kWh] is multiplied by the current electricity price [Price/kWh]. You can find the current electricity price on the internet.

 

To measure amperage and voltage, two electronic modules are used, which are built into the electric vehicle charging station.

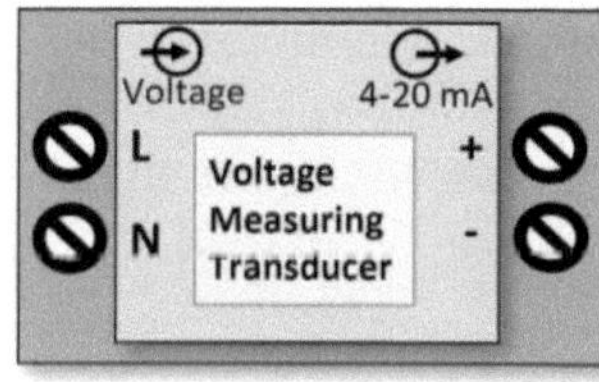


**Voltage Measuring Transducer**
This electronic module measures the current voltage from the power supply, for example, it can be 230 V. The current voltage is an analog value that cannot be directly connected to an analog input module on a PLC. Therefore, an electronic module such as a voltage measuring transducer is required. The module performs a linear scaling (conversion) of a voltage in the range of 0 to 400 V to a value in the range of 4-20 mA.

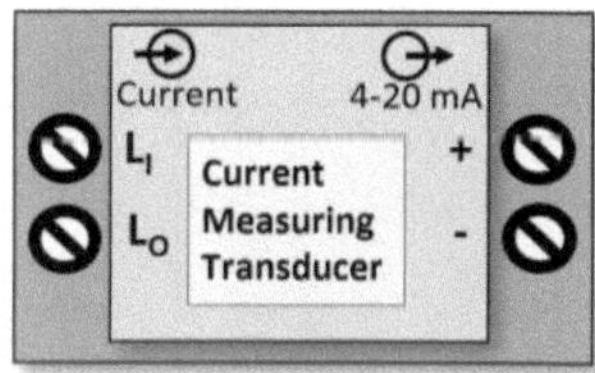


**Current Measuring Transducer**
This electronic module measures the battery charging current, for example, it can be 20 A. The current is an analog value that cannot be directly connected to an analog input module on a PLC. Therefore, an electronic module such as a current measuring transducer is required. The module performs a linear scaling (conversion) of a current in the range of 0–40 A to a value in the range of 4–20 mA.

**Calculation**
The current consumption (power) P is measured in watts [W] and is calculated as follows:

$P = U \cdot I \cdot \cos(phi)$, where U = voltage, I = current, cos(phi) is set to 1.

The current consumption is usually displayed in kilowatts [kW], and kW = W/1000.

The amount used is measured over a period of time and is displayed in kWh (kilowatts per hour).

**Example calculation**
If an electric car is charging with 10 A, the voltage is 230 V, and it charges for 15 minutes, the consumption is:

$(10 \cdot 230 \cdot 15)/(60 \cdot 1000) = 0.575$ [kWh], where 60 represents the number of minutes in 1 hour.

Remember that the current consumption (power) is not constant during the charging of a battery, which means an average of a series of measurements should be taken. You must determine an appropriate average.

**Exercise**
Write a PLC program that can calculate the cost of a charging session.

# 7 Applied mathematics in PLC programming

This chapter contains exercises where relevant applied mathematical formulas and calculations need to be programmed in a PLC.

A lot of PLC programs require the use of mathematics and formulas; therefore, the exercises in this chapter are a good starting point for learning this.

## 7.1 Calculate the volume of a process tank

A process tank is shaped like a cylinder.

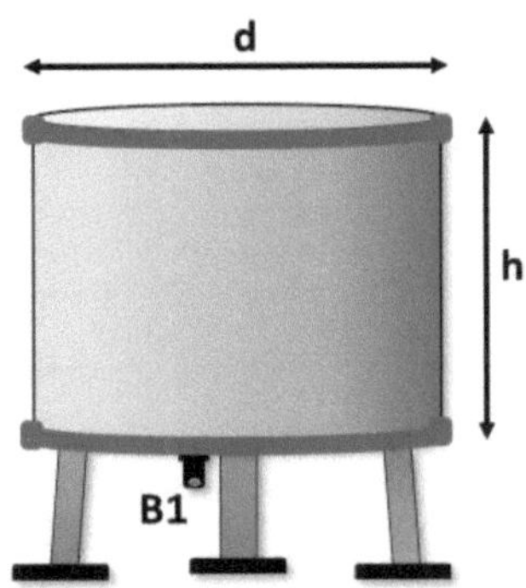


**Data for the process tank**
Diameter: d = 1.4 meters
Height: h = 2.75 meters

An analog level sensor **B1** is mounted at the bottom of the process tank; it measures the liquid level.

**Exercise**
Write a PLC program that calculates the volume in the process tank based on the current measured liquid level.

## 7.2 Calculate the volume of a tall silo

A tall silo is shaped like a cylinder with a truncated cone at the bottom.

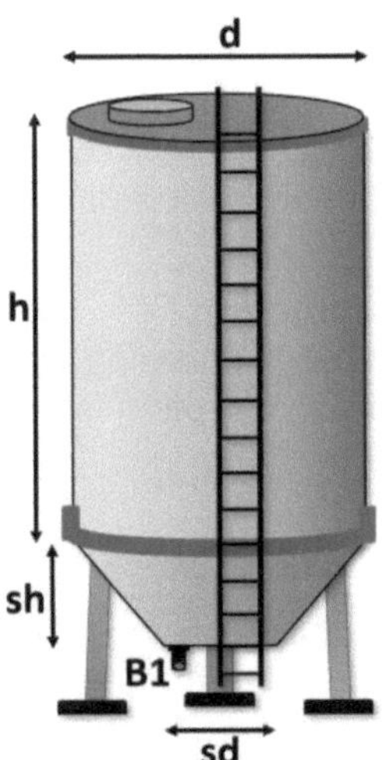


**Data for the silo**
Diameter: d = 3.4 meters
Height: h = 8.50 meters

Truncated cone height: sh = 1.2 meters
Truncated cone diameter: sd = 0.8 meters

An analog level sensor **B1** is mounted at the bottom of the silo, measuring the liquid level.

**Exercise**
Write a PLC program that calculates the volume in the silo based on the current measured liquid level.

 

## 7.3 Calculate the speed of a conveyor belt

Two digital sensors **B1** and **B2** are mounted on a conveyor belt.

The distance between the two sensors is d = 0.8 meters.

**Exercise**

Write a PLC program that calculates the speed of the conveyor belt.

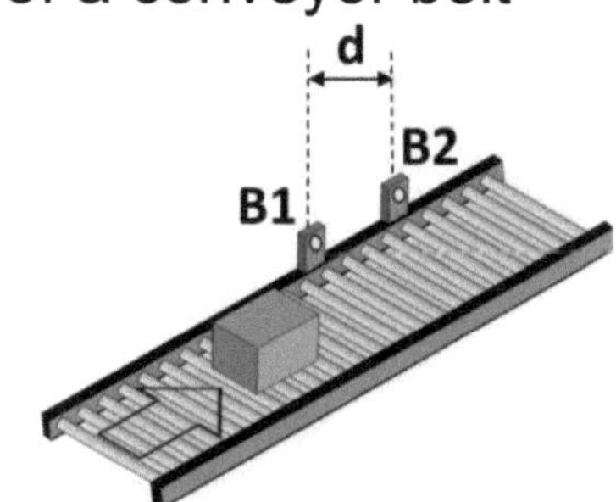


## 7.4 Calculate when the rainwater basin is full

A large rainwater collection basin looks like this:

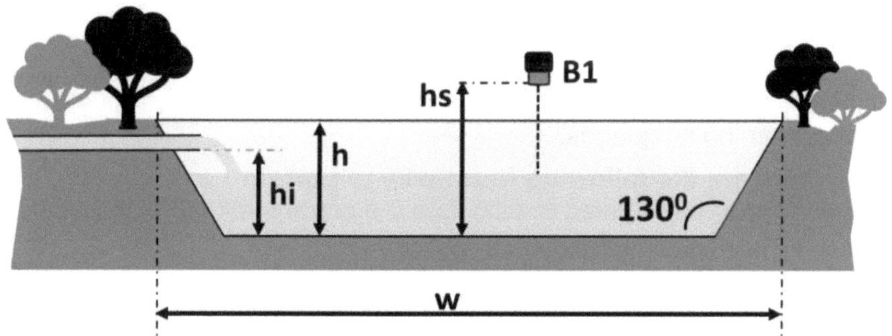


The basin is shaped like an inverted truncated pyramid, with the widths being equal.

**Data for the basin:**
Width: w = 12 meters
Height of the basin: h = 4.00 meters
Height to inlet pipe: hi = 3.50 meters
Height to level sensor **B1**: hs = 5 meters

An analog sensor **B1** measures the distance down to the water level.

The basin is filled with water when the water level reaches the inlet pipe (measured water level = hi).

**Exercise**

Write a PLC program that continuously calculates the time it takes for the basin to be filled with water.

 

## 7.5 Calculate the volume of a horizontal cylindrical tank

A company develops and sells large tanks for storage liquids. The tanks are cylindrical in shape and are placed horizontally. The ends of the tanks are flat.

Here are images of one tank:

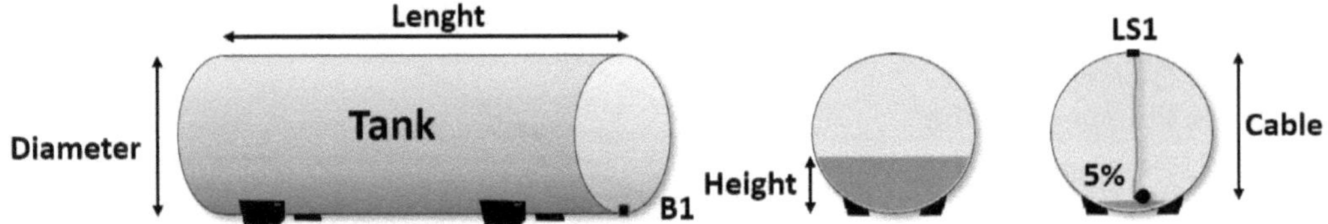


**Description**

The company would like to sell two sensors together with the tank:

- Sensor **B1** for measuring the level in the tank (analog level measurement).
- Low-level switch **LS1** (low-level contact or float switch).

Furthermore, the company would like to give away the necessary PLC program to calculate the current liquid volume in the tank, allowing customers to easily integrate the tank calculations into their own PLC systems. Customers use various PLC brands and PLC programming languages.

**Due to this, there are the following four tasks to perform:**

**1)** Write the PLC program needed to calculate the current volume in the tank.

Remember to explain why you chose the solution(s).

**Module test**

To test your PLC program, a tank with the following data is selected:

Diameter of 4 meters and a length of 6 meters.

**2)** Design and complete a test table/form with the necessary number of test points. This means selecting an appropriate number of liquid levels in the tank and the expected calculated volume. Then test your PLC program where you write down the calculated results from the PLC program in the test table form. Then use the test table/form to verify that the PLC program calculates correctly.

**3)** The low-level switch **LS1** should give a digital signal when there is only 5% liquid (volume) left in the tank.

How far should the sensor cable be inside the tank when the sensor hangs down from the top of the tank?

**4)** Describe and demonstrate how it is possible to program a solution that continuously monitors whether the two sensors are working correctly.

 

## 7.6 Calculate the quantity of stones and soil

In this exercise, you must develop a PLC program that can calculate the quantity of stones and soil coming from a conveyor belt in a gravel pit.

Illustrations:

**Functionality**

A conveyor belt moves stones, gravel, soil, or sand to create a large pile. The conveyor belt has two analog distance sensors, **B1** and **B2**, installed.

The three illustrations show the following:

❶ Initial situation. There is no pile, and the two sensors measure the same distance.
❷ This pile consists of both stones and soil, making the pile very tall and narrow.
❸ This pile consists only of soil and is therefore flatter than it is tall.

The distance between the two sensors is d = 50 cm.

**The shape of each pile**

It is supposed that the pile of materials has a shape resembling a truncated cone but with differing slopes, as piles of stones, gravel, soil, or sand have varying slopes. Depending on the material, some piles are tall and narrow, while others are flatter.

**Exercise**

Write a PLC program that continuously calculates the following:

1) The quantity (volume) of material in the pile.
2) The quantity (volume per unit time) of material added to the pile.
3) Continuously calculate the point in time when there will be no more space for material under the conveyor belt.

## 7.7 Distance between wooden planks on conveyor belt

A machine cuts a long wooden plank into smaller ones, and the planks are placed close to each other on the conveyor belt **M1**.

The planks need to be spaced at a certain distance. This is achieved by ensuring that there is a higher speed on conveyor belt **M2** than on conveyor belt **M1**, as dictated by the PLC program.

Illustration:

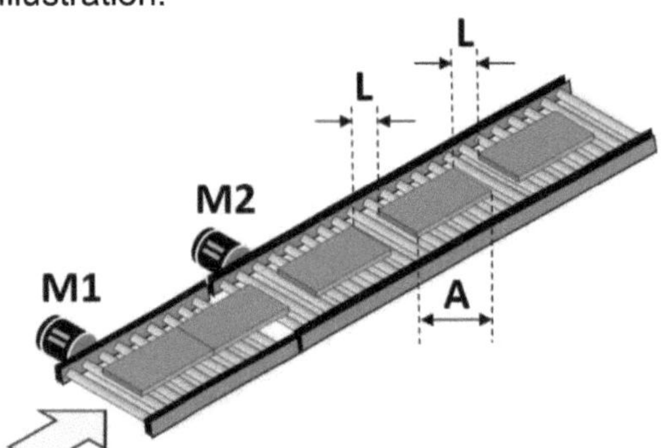


**Description**

The distance between the plates is **L**, and the plates have a size **A**.

The two motors, **M1** and **M2**, which drive the two conveyor belts, are identical.

The diameters of the rollers on the two conveyor belts are identical as well.

The motors **M1** and **M2** are controlled by an analog 4–20 mA signal from the PLC, and the speed of the motors can be adjusted to be between 0 and 1000 RPM (revolutions per minute).

**Exercise**

Write a PLC program that can set the RPM speed of motor **M2** based on the RPM speed of motor **M1**, so the distance between the wooden plates is **L**.

## 7.8 Fill concrete truck

In this exercise, you must write a PLC program to calculate the amount of filling a concrete truck (cement truck).

**Description**:

A conveyor belt is controlled by a motor, **M1**. It has the shape of an isosceles trapezoid.

The conveyor belt is used to fill a concrete truck with liquid concrete. The height of the concrete layer on the conveyor belt is measured by an analog sensor, **B1**. The sensor measures the distance (Lm) down to the concrete layer. This distance can naturally vary while the conveyor belt is in operation and the concrete truck is being filled with liquid concrete.

 

The setup is as shown below:

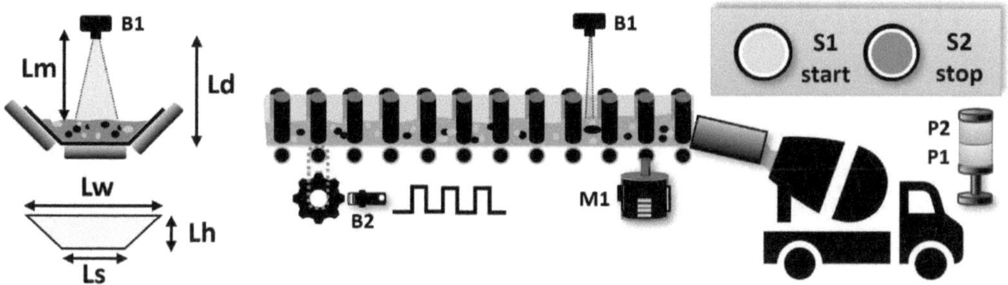


The conveyor belt is equipped with an encoder sensor **B2**, which gives digital pulses when the conveyor belt is in operation.

There can be up to 12 $m^3$ of liquid concrete in a concrete truck.

**Description of components:**

| Name | Type | Description |
|---|---|---|
| **M1** | Motor | Drives conveyor belt. Set to **TRUE** for the conveyor belt to operate. |
| **B1** | Sensor | Measures distance down to the liquid concrete.<br>Measurement range: 0 to 2 meters. |
| **B2** | Sensor | Sensor used as an encoder. Digital signal. Roller diameter: 10 cm. |
| **S1** | Push button | (NO) Start operation. Concrete truck is filled with the selected amount of concrete. |
| **S2** | Push button | (NC) Stops immediately. Stops conveyor belt. |
| **P1** | Lamp | Light turned on when conveyor belt is in operation. |
| **P2** | Lamp | Light turned on when conveyor belt is not in operation. |

**Table of distances:**

| Distance | Description |
|---|---|
| **Lm** | Distance to the top of the concrete. Measured by sensor **B1**. |
| **Lw** | Maximum width of the conveyor belt: 80 cm. |
| **Ls** | Minimum width of the conveyor belt: 40 cm. |
| **Lh** | Height of the conveyor belt: 30 cm. |
| **Ld** | Distance from sensor to bottom of the conveyor belt: 140 cm. |

It should be possible for the operator to select the amount of concrete to be filled into the concrete truck. This selection cannot be changed after the operator has pressed the start button. The system will stop automatically when the concrete truck is filled with the chosen amount of concrete.

**Exercise**
Write a PLC program that can fill the concrete truck.

# 8 Use of ARRAY in PLC program

This chapter includes exercises where you can learn to use ARRAY.

## 8.1 Traffic sign trailer for road control

In this exercise, you must develop a PLC program for a traffic sign trailer that can be used to regulate traffic on the road. Illustration of the traffic sign trailer:

**Description**

The traffic sign trailer has a frame with lamps on the back that can be moved up when traffic needs to be regulated. The lamps are mounted in a matrix on the frame, and the lamps can be turned on to display the following signals to the drivers of vehicles:

**Lamp signals**

The signals mean:

1) Turn left.
2) Lane closed.
3) Turn right.
4) Maximum speed 30.
5) Maximum speed 50.
6) Lights turned off.

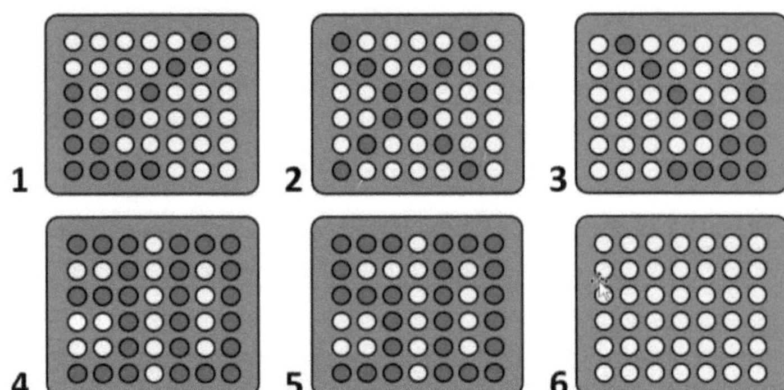


**Control panel**

There is a control panel where the operator can press an image to select a signal.

In this exercise, you must define appropriate names for variables (TAGS) yourself.

**Exercise**

Write a PLC program for the traffic sign trailer.

 

## 8.2 Sorting data from a machine

A machine places four bottles into a crate:

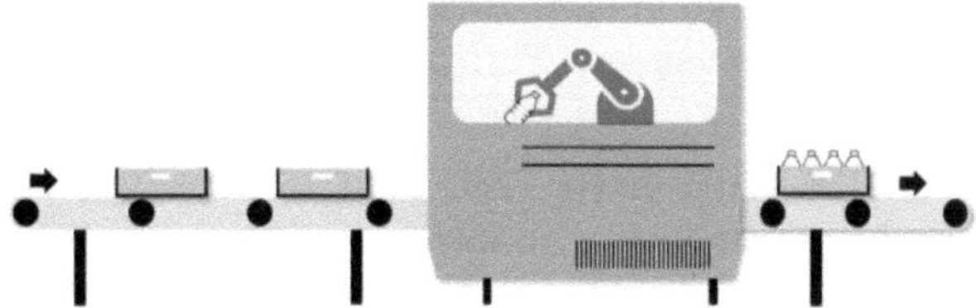

**Description**

A PLC counts the number of crates that have passed through the machine each hour.

To have a good program structure, the number of crates is collected in an ARRAY.

Here is an ARRAY containing the number of crates that have passed through the machine each hour for the last 12 hours:

| 121 | 47 | 67 | 165 | 119 | 82 | 0 | 145 | 114 | 78 | 67 | 45 |
|---|---|---|---|---|---|---|---|---|---|---|---|

**Exercises**

You are now to solve five exercises related to the collected data:

**1) Create ARRAY**

You must create an ARRAY in the PLC program with the collected data.

**2) Total number of crates**

You must now code a loop that can be used to count the total number of crates that have passed through the machine in the last 12 hours.

**3) Average**

Using a loop code, you must find the average number of crates per hour that have passed through the machine.

**4) Best production**

Using a loop code, you must find the highest number (165) that has passed through the machine.

**5) Sorting data**

You must now sort the data to provide (give) a better overview of the collected data. The ARRAY should be sorted so that the largest values are placed to the left and the smallest values to the right, as shown below:

| 165 | 145 | 121 | 119 | 114 | 82 | 78 | 67 | 67 | 47 | 45 | 0 |
|---|---|---|---|---|---|---|---|---|---|---|---|

# 9 Development of functions and function blocks

This chapter includes exercises to learn designing and programming of functions and function blocks for use and reuse in a PLC program.

## 9.1 Function that can add two numbers together

Create a function that has two input variables: **Number1** and **Number2**.
The function should add the two input variables together and return the result.
The name for the return variable should be **Sum12**.
Find a suitable name for the function yourself.

Block diagram:

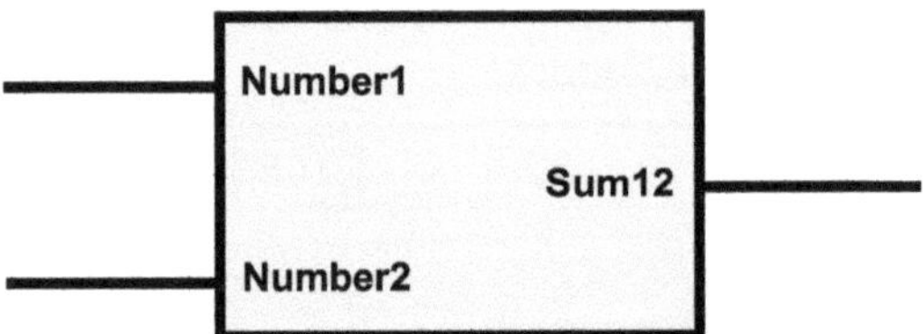


## 9.2 Function that can multiply two numbers together

Create another function that has two input variables: **Number3** and **Number4**.
The function should multiply the two variables together and return the result.
Unlike the previous exercise, the result should not be placed in a separate variable, but should be returned via the function's built-in return value.

Find a suitable name for the function yourself.

Block diagram:

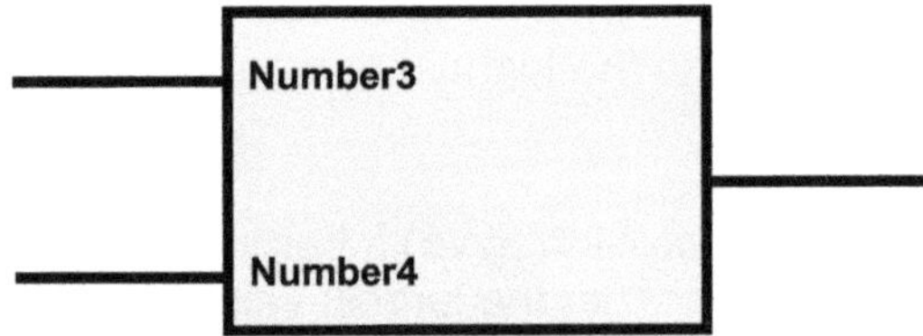


 

## 9.3 Function block that can add two numbers together

Create a copy of the function from the previous exercise, where two numbers were added together, but now it should be a function block.

**Counter**

Add a counter inside the function block that increments by 1 each time the function is called. The counter value should also be returned, so the function now has two return variables: the sum of the numbers and the counter value.

Find a suitable name for the function yourself.

Variable names should be as shown in the block diagram below:

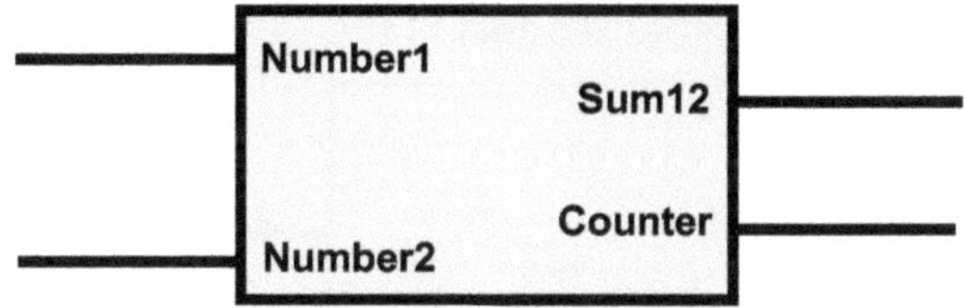


## 9.4 Function that makes function calls to other functions

A function should return a result where two numbers are either added together or multiplied together.

You need to use the functions created in exercises 9.1 and 9.2.

**Input variables**

Create a function that has three input variables and one return value: Two of the variables are numbers; they are named **Number5** and **Number6**, and the third is a variable that determines what should happen with the two numbers.

The third variable should be able to have two different values: if it is one value, the function created in exercise 9.1 should be called, so the two numbers are added together; otherwise, the function created in exercise 9.2 should be called, so the two numbers are multiplied together.

You must define the data type for the variable **FuncSelect** and the two different values **FuncSelect** that can be chosen to trigger a function call to one of the two functions.

Block diagram of the function:

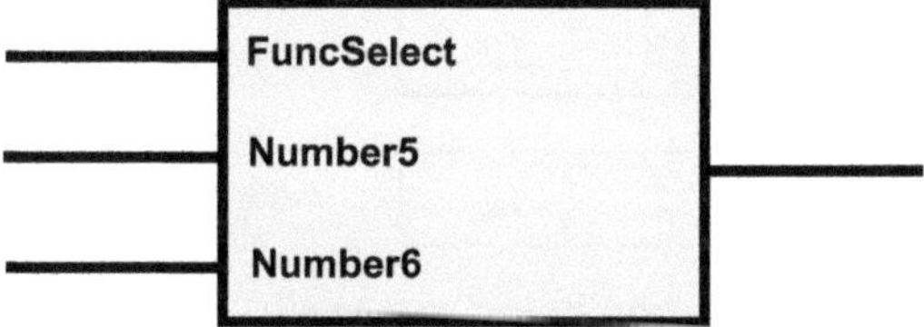


The function must return the calculation of **Number5** and **Number6**.

## 9.5 Function for calculating speed of a box

In this exercise, you must write a function block that calculates the speed at which a box is moving on a conveyor belt between sensor **B1** and sensor **B2**.

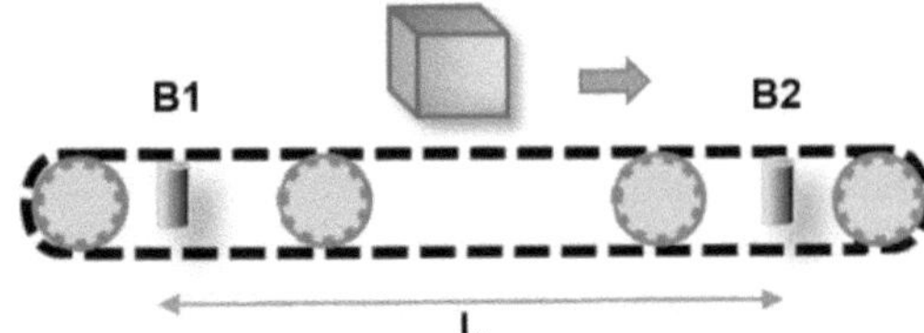


Sensors **B1** and **B2** give a **TRUE** signal when the box is in front of the sensor.

Design proposal for a function block where **B1**, **B2**, and the distance **L** are input values to the function block and the variable **Speed** is the output value:

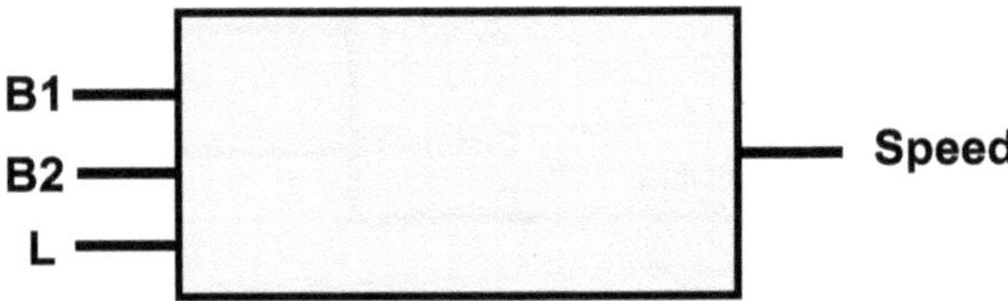


## 9.6 Function block for averaging measurement values

Program a function block that calculates the running average of the three most recent measurements from an analog sensor. The function block should have one input value (the measurement) and one output value (the average).
Here is the calculation of the average for five function block calls, where the most recent values from the sensor are 10, 12, 11, and 13:

| | | | | |
|---|---|---|---|---|
| 1. | 0 | 0 | 0 | = 0 |
| 2. | 10 | 0 | 0 | = 3.3 |
| 3. | 12 | 10 | 0 | = 7.3 |
| 4. | 11 | 12 | 10 | = 11 |
| 5. | 13 | 11 | 12 | = 12 |

The numbers on the far right show the calculated average.

 

## 9.7 Function block as a 24-hour clock

Program a function block that serves as a 24-hour clock.
It should be usable on all types of PLCs.
The time readout should be in seconds (ss), minutes (mm) and hours (hh).

Proposal for the design of the function block:

It should be possible to set the clock to the current time.

## 9.8 Switching between a 24-hour and a 12-hour clock

Design and program a function that can switch between time formats of a 12-hour clock or a 24-hour clock. Design proposal:

## 9.9 Conversion between hours and decimal hours

Usually, time is displayed in hours and minutes. If two time periods need to be added together, it can be advantageous to convert hours and minutes to display in decimal hours. For example, 10 hours and 30 minutes should be represented as 10.5 in decimal hours.
Program a function that converts from hours to decimal hours and a function that converts from decimal hours to hours.

Design proposal:

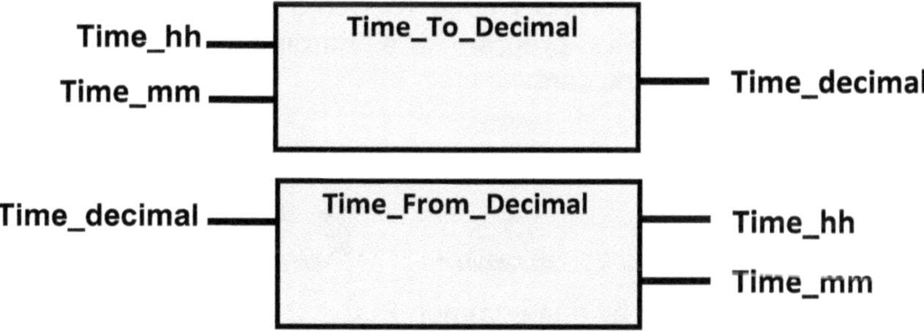

## 9.10 Calculation the volume of a box

To measure the volume of a box, three analog sensors and the speed of a conveyor belt are used.
Boxes are always square and placed parallel on the belt.

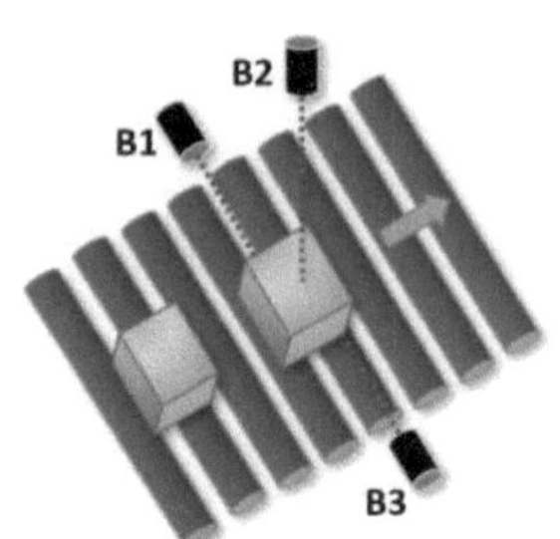


Design and program a function or function block that can calculate the volume of a box.

## 9.11 Calculating the number of pallets in a pile

Design and program a function that can count (measure) the number of pallets stacked in a pile.

All pallets in the same stack are of the same type.

A distance sensor **B1** needs to be used, and it is placed above the pallets.

The function should have the pallet size **PalletSize** as an input variable so that it can be easily reused for multiple types of pallets.

The return variable can be **PalletCount** and it contains the number of pallets.

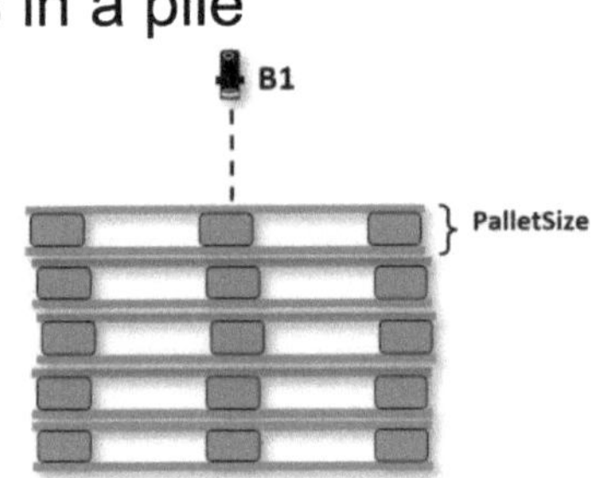


Proposed design for the function:

## 9.12 Charging station for five electric cars

In this exercise, you must write a PLC program for a charging station with the capacity to charge up to five electric cars:

The charging station is controlled by one common PLC.

 

**Customer login**

When an electric car needs to be charged, the customer must first log in with their customer number and PIN code. The customers are created in the PLC program, and they are shown in the table to the right.

Once the customer is logged in, they should choose one of the five charging stations and press "start" to initiate a charging session.

| Customer number | PIN code |
|---|---|
| 00001 | 2356 |
| 00002 | 9080 |
| 00003 | 1356 |
| 00004 | 6532 |
| 00005 | 9012 |
| 00006 | 5490 |

**Energy measurement**

An energy meter used for measuring charging energy consumption:

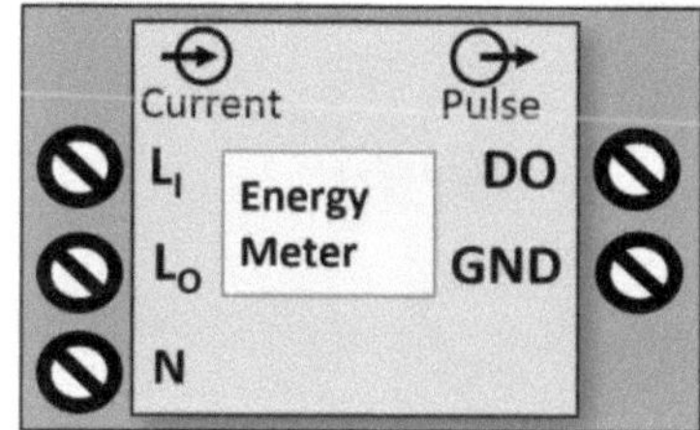


**Energy Meter:**

In an electrical panel within each charging station, an energy meter is installed to measure the current consumption during the electric car charging.

The energy meter has a digital output (DO) connected to a digital input on a shared PLC. There is a total of five energy meters.

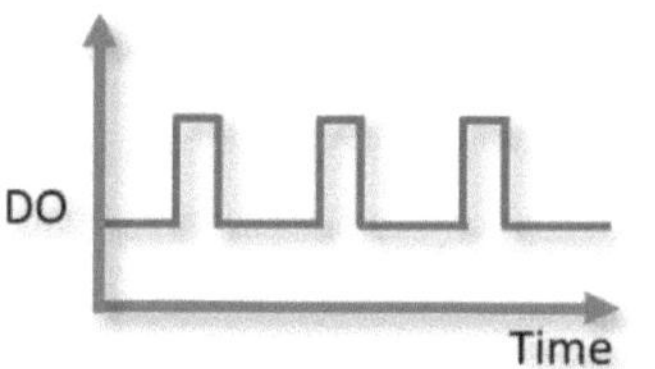


**Pulse train:**

Every time an energy meter measures 1 [Wh], a pulse is generated on the digital output DO, and the pulse duration time is 150 ms. This means that for every 1000 pulses, 1 [kWh] is consumed. It is thus the common PLC that collects all the pulses for a charging session and calculates the total consumption.

**Calculation of price**

To calculate the price to be paid for the charging, the amount used [kWh] is multiplied by the current energy price. You can find the current energy price [kWh] on the internet.

When the customer presses the "stop" button, the charging is stopped, and the price is displayed until the next customer logs in.

You need to design the user interface yourself.

**Exercise**

Design and write a PLC program for the charging station.

# 10 Programming with STRING and STRUCT

This chapter contains exercises where you learn to use **STRING** in a PLC program. Further, there are exercises with **STRUCT** and **ENUM** (enumerate).

## 10.1 Alarm codes from a frequency converter (VDF)

A frequency converter is connected to a PLC with a fieldbus, as shown:

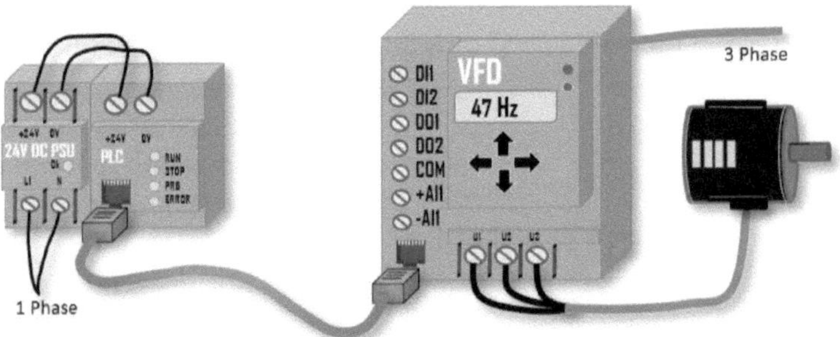


**Description**

The frequency converter (Variable Frequency Drive – VFD) has inside alarm monitoring for a wide range of values and measurements. Each alarm has an alarm code (a number), and all alarm codes are collected in the frequency converter and sent to the PLC in a **STRING**.

The content of a **STRING** may look like this:

"120;34;30;1;45;30;12;10;90;30;130;10"

There are 12 alarm codes in the example shown, but there may be a different numbers of alarm codes in each **STRING**. The alarm codes are separated by semicolons, and as can be seen, a single alarm code may occur multiple times.

**Exercise**

Write a program that counts the number of times the alarm code '30' occurs.

The solution to the exercise should include two proposed solutions:

**A)** The built-in standard **FIND** function should be used to find the number of times the alarm code '30' occurs.

**B)** Alarm codes are moved to an **ARRAY**, and a **FOR** loop is used to iterate through the entire **ARRAY** and count the number of times the alarm code '30' occurs.

 

## 10.2 Design and print label for chicken

In this exercise, you must develop a program that can print a label for a chicken. The chicken is placed in a plastic tray ❶ and transported on a conveyor belt to a scale ❷ to be weighed. Then the chicken must have a label printed by a printer ❸:

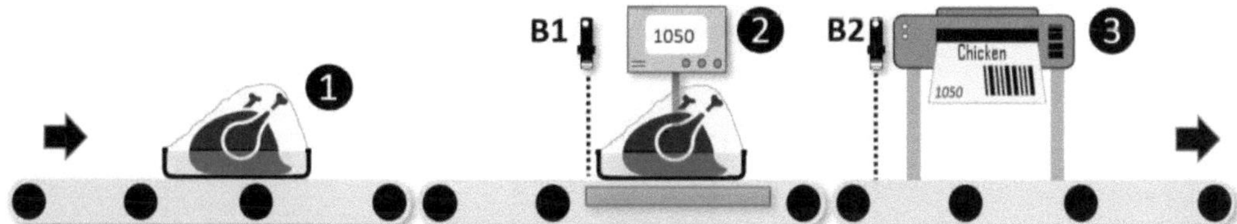


**The Scale ❷**
The chicken is ready to be weighed when the signal from digital sensor **B1** changes from **TRUE** to **FALSE**.

The scale has a 4-20 mA analog output, indicating the current weight.

Measurement range: 0 to 2000 grams.

**Label Printer ❸**
The label printer is connected to the PLC via a fieldbus.

The following variables are available in the fieldbus protocol for the label printer:

| Variable name (TAG) | Data type | Description |
|---|---|---|
| **TextLine1** | **STRING** | Max. 40 characters. |
| **TextLine2** | **STRING** | Max. 40 characters. |
| **TextLine3** | **STRING** | Max. 40 characters. |
| **TextLine4** | **STRING** | Max. 40 characters. |
| **WriteText** | **BOOL** | Print on a positive edge. |

The variables **TextLine1** to **TextLine4** must be filled with the text to be placed on the label, and the data should be sent to the printer when the variable **WriteText** is set to **TRUE**. The printer prints a label on a positive edge of the **WriteText** variable. The label should be printed when the chicken is at the printer. The chicken is at the printer when the signal from the sensor **B2** changes from **TRUE** to **FALSE**.

**Minimum requirements for the text on label (Max. four-line text):**

- The current weight of the chicken.
- The packing date (the current day and time).
- Best before date: 10 days from the current date.

**Exercise**
Design a label for the chicken and write a PLC program for printing.

## 10.3 Sorting of packages using vision camera

In this exercise, you must write a PLC program for sorting packages.
The setup is shown below:

**Description**

The packages are transported on a conveyor belt (8). A motor (actuator) with a moving arm (1–4) can swing out, and a metal chute ensures that the package goes into the correct container. Each package (9) has a label with a postal code (ZIP code) that can be in a range from 0000 to 9999.

**Vision camera**

A vision camera (5) can directly read the postal code on the package, as shown in the three images below:

The vision camera is connected to the PLC with a network cable (fieldbus), and therefore, the postal code can be directly read in the PLC program from a variable, which has the data type **STRING**.

If the vision camera cannot see a postal code on a package, the **STRING** variable is empty (""). This means that the **STRING** variable is set to "".

A package with the postal code "PC2035" should go into container number 2, as this container contains packages with a postal code in the range of 2000 to 4999.

Packages without an address label, without a postal code, and with an unknown postal code should go into the last container (6), which is located at the end of the conveyor belt. Postal codes such as "AB00" and "0000" are thus invalid values.

There is only one package on the conveyor belt at a time. It takes 10 seconds from the vision camera reading the postal code to the package being in the correct container. After this, the moving arm must be reset so the next package can pass by.

**Exercise**

Write a PLC program that can sort the packages into the correct container.

 

## 10.4 ENUM for a motor (OperatingState)

A motor can be in these various operating states:

Stop, Starting, Run, Stopping, Alarm, Service, None, Manual

These are collected in an **ENUM** which should be named **OperatingState**.

The state **None** is set as the default value.

Note that not all PLC manufacturers and PLC brands support the use of **ENUM**. If **ENUM** is not possible in the PLC you are using, you can use **CONST**.

**Exercise**
Create an **ENUM** according to the mentioned requirements.

## 10.5 STRUCT for a motor (MotorType)

A **STRUCT** named **MotorType** should contain the values:

TAGName, MotorID, Power, **OperatingState**, Temperature, Hz, AlarmBit

The value **OperatingState** was defined in the previous exercise (10.4).

You must find the appropriate data types and select useful default values yourself.

**Exercise**
Create a **STRUCT** according to the mentioned requirements.

## 10.6 Startup and operation of 10 motors

A small plant has 10 motors:

Create an **ARRAY** with 10 motors of type **MotorType**. (Created in exercise 10.5).

**Exercise**
Write a PLC program that starts all 10 motors.

First, the motor should be set to the **Starting** state for 2 seconds, then it should be set to the **Run** state. There should be a 1-second delay between each time a motor is set to the **Starting** state, so the motors do not start at the same time and overload the power supply.

A motor must not go into the **Starting** or **Run** state if the **AlarmBit** is **TRUE**. A motor must stop immediately if the **AlarmBit** becomes **TRUE**.

A motor that has stopped must not start again until all 10 motors have stopped and are started again.

# 11 Various complex exercises

This chapter contains several types of programming exercises for a PLC. Many of the exercises are complex and require a lot of experience, making them most suitable for the experienced PLC programmer who seeks more challenges. Some of the exercises require specification, component selection and detail program design.

## 11.1 Speed measurement of trucks

In the company where you are on an internship, there are many trucks that are driving too fast. Therefore, you are assigned a task to write a PLC program that can measure the current speed of each truck and count the number of trucks that are speeding.

**Description**

The current speed is measured using two sensors, and the measured speed should be displayed on a large display. Below is an illustration:

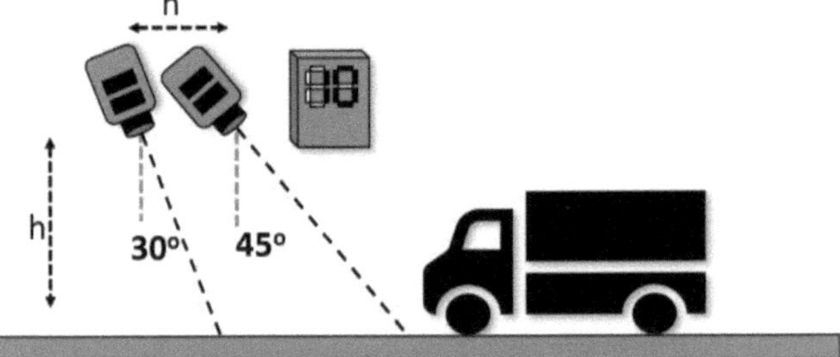


The distance from the truck to the two sensors is the height **h**, where **h** is 4 meters.

The distance between the two sensors is **n**, where **n** is 0.5 meters.

The two sensors are positioned at angles of 30 degrees and 45 degrees, respectively, relative to the vertical.

Each sensor gives a **TRUE** signal when the front of the truck is within the sensor's measurement range. This is indicated by a dashed line in the illustration above.

**Displaying Speed**

To display how fast a truck is driving, a large display is used. This display consists of two 7-segment displays.

A 7-segment display can show numbers from 0 to 9 using seven lamps (LED light):

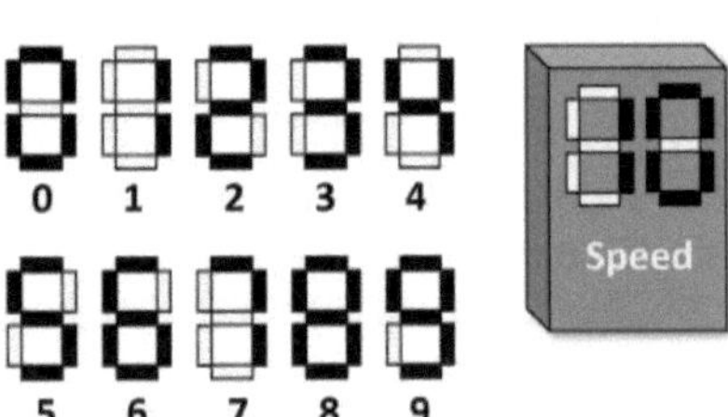


If all lamps in a 7-segment display are turned on, the number 8 is displayed.

When two 7-segment displays are used, numbers between 0 and 99 can be displayed. The example above shows the number 10.

 

Further, there is a small control panel that contains the following:

On Off

Manual switch for turning on and off the control and the large display.

Manual switch used to toggle between displaying the speed in either km/h or mph.

Analog rotary potentiometer used to set the maximum speed a truck may drive.

When the potentiometer is turned, the maximum speed is displayed on the display, allowing the user to see the maximum speed set. After 10 seconds, the display turns off, but the maximum speed is still stored in the PLC program.

A value between 0 and 99 can be set.

If the truck is driving faster than the maximum speed, both digits in the seven-segment display should blink, alerting the truck driver that they are driving too fast.

Every time a truck has driven too fast, an internal counter in the PLC program should register this, allowing the number of trucks that have driven too fast to be displayed.

This is a manual push-button used to display the number of trucks that have driven too fast.

When the button is pressed, the number of trucks that have driven too fast is displayed. The number is shown on the large display.

If the button is pressed for 10 seconds, the internal truck counter is reset to 0.

If the number of trucks that have driven too fast exceeds 99, 99 is displayed.

In this exercise, you must define useful names for variables (TAGS) yourself.

**Exercise**
Write a PLC program based on the description.

## 11.2 Controlling of pump speed using PWM

The speed of a pump **M1** needs to be controlled from a PLC. This is done by using a digital signal from the PLC, as the pump does not have an analog input or the option for a fieldbus connection. The digital signal is a low-cost method for enabling a PLC to control a pump. The pump is a circulation pump used for pumping water, for example, in heating installations in buildings.

The method used to control the speed of the pump is called PWM (pulse width modulation). This means that the pulse width of the digital signal is changed when the speed needs to be adjusted.

This is illustrated below with timing diagrams ❶ to ❺:

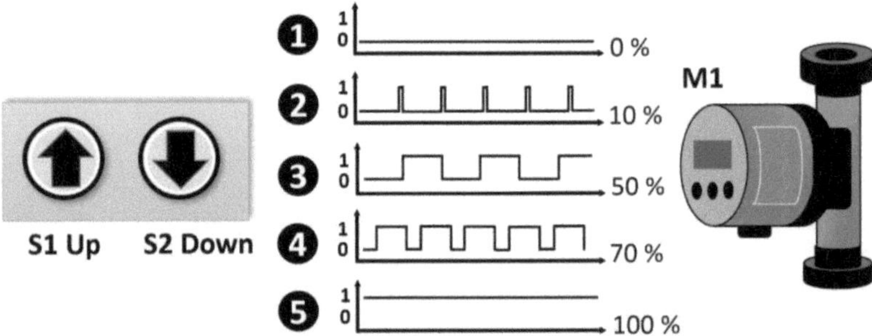


The timing diagram above shows:

| | |
|---|---|
| ❶ | If the digital signal from the PLC is 0 (**FALSE**) all the time, the pump is stopped. |
| ❷ | Here, the signal is 1 (**TRUE**) for 10% of the time. This means the pump is running at 10% of the maximum speed. |
| ❸ | Here, the pump is running at half the maximum speed. |
| ❹ | Here, the pump is running at 70% of the maximum speed. |
| ❺ | If the digital signal is 1 (**TRUE**) all the time, the pump is running at maximum speed. |

**User interface**

To change the speed, the PLC must use two manual push buttons, **S1** and **S2**. When the push button **S1** is pressed, the speed should increase, eventually reaching the maximum at 100%. When **S2** is pressed, the speed should decrease, eventually reaching 0%, when the pump is completely off.

**Exercise**

Write a PLC program based on the description.

 

## 11.3 Conveyor belt with yielding

In this exercise, you must develop a PLC program that can manage the control of three conveyor belts.
There is a main conveyor belt, **M1** and **M2**, as well as a side conveyor belt, **M3**.

Illustration:

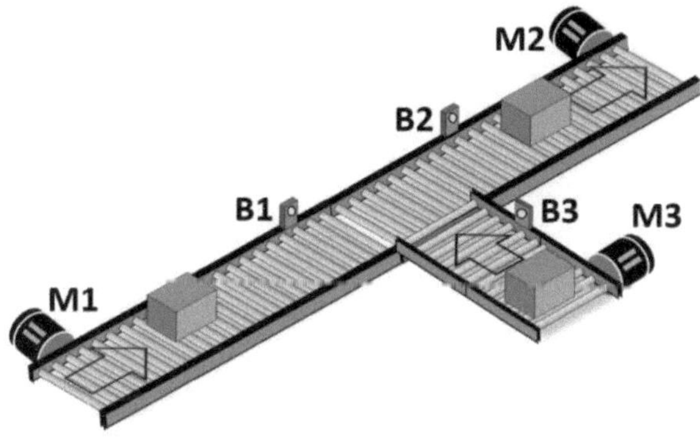


**Description**

Each conveyor belt is controlled by a motor, which runs when it receives a digital **TRUE** signal from the PLC, and stops when the digital signal is **FALSE**.
Each conveyor belt is equipped with a sensor that gives a **TRUE** signal when there is a package in front of it. When there is no package in front of the sensor, it gives a **FALSE** signal.
All three motors are set to run at the same speed.

**Exercise**

The packages from the side conveyor belt **M3** should be transferred onto the main conveyor belt without colliding with a package that is moving from conveyor belt **M1** to conveyor belt **M2**.

Write a control specification describing how the PLC control should operate. It should include the necessary descriptions, calculations, and information from which a PLC program can be directly written.
After this, then write a PLC program based on your control specification.

## 11.4 Centrifuge with ramp control of the speed

A centrifuge needs to be controlled from a PLC. A centrifuge is a circular container that rotates rapidly and is often used in industries to separate liquids. It works similarly to the way a washing machine centrifuges to remove water from laundry.

Illustration:

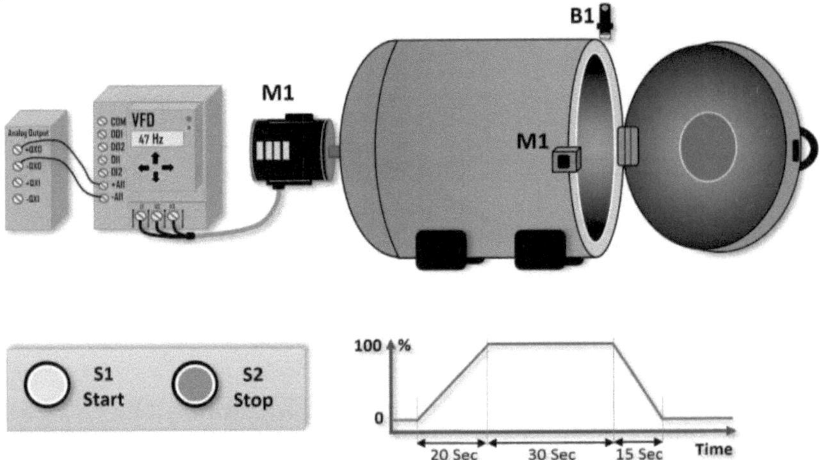


**Description**

A large motor **M1** is used to rotate the centrifuge. The motor is controlled via a frequency converter (Variable Frequency Drive – VFD), and the frequency converter is controlled with an analog signal.

Due to the size of the centrifuge, it is necessary to use ramp control, which ensures that the rotation speed starts and stops slowly.

It takes 20 seconds for the rotation to reach full speed. It runs at full speed for 30 seconds and then ramps down for 15 seconds.

The centrifuge must only start when the door switch contact **B1** gives a **TRUE** signal and the door lock **M1** is set to **TRUE**.

When the centrifuge has stopped, the door switch contact is set to **FALSE** so the door can be opened.

**User interface**

To start the centrifuge, the push button **S1** is used. The centrifuge stops automatically after 65 seconds.

At any time, the centrifuge can be stopped with push button **S2**. It should ramp down if **S2** is pressed because a centrifuge can break if it stops right away.

**Exercise**

Write a PLC program based on the description.

 

## 11.5 Parking garage with vision camera

A modern parking garage uses the vehicle's license plate to calculate the parking fee.

A license plate is the same as number plate (Car ID number).

**Description**

The system, known as automatic number plate recognition (ANPR), is a modern and user-friendly alternative to traditional parking barrier systems. It provides the same benefits as a traditional system, but replaces barriers with cameras. This allows drivers to enter and exit the garage without needing to wait at a barrier for a ticket.

Illustration of a typical parking garage:

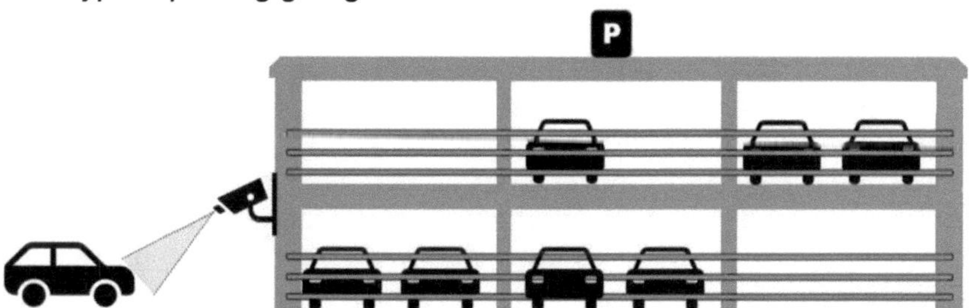


When entering the parking garage, the license plate is scanned by a camera. Afterward, the driver finds a free parking spot in the garage.
Before leaving the parking garage, the driver must pay for parking at a ticket machine.

**Payment**

The first 30 minutes of parking are free. After that, it costs 0.5 EUR/US$ per minute, with a minimum charge of 1 EUR/US$. The time starts from the moment the camera has scanned the license plate of the entering vehicle. The time ends when the driver has paid for parking. Once paid, the parking garage must be exited within 15 minutes. There is a camera at the exit that registers when the driver has left the parking garage.
If the driver has not paid after 24 hours, a fee of 100 EUR/US$ must be paid.
The parking garage has 100 parking spots.
The license plate is scanned by a camera connected to the PLC via a fieldbus. The text from the license plate is stored in a **STRING** variable.
There is an HMI on the ticket machine where the driver can pay for parking.

**Exercise**

Develop a program design and propose an HMI (user interface for ticket machine). Then write a PLC program for the parking garage.

## 11.6 Collection of wastewater from a factory

This setup consists of pallet tanks used for collecting wastewater from a factory. The setup is as shown below:

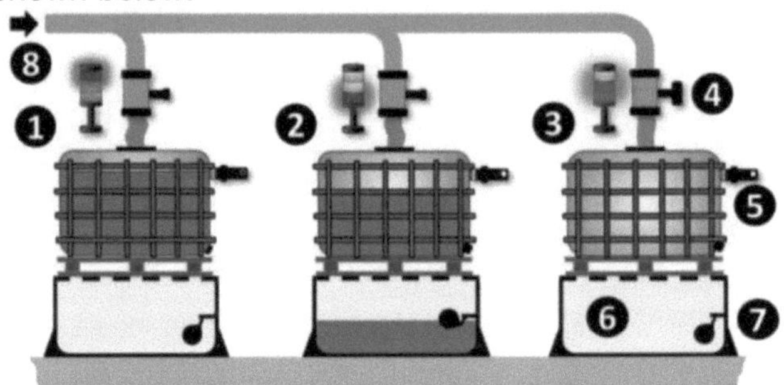


**Description**

The wastewater is collected in pallet tanks so that it can be transported to a treatment plant, which can properly and environmentally clean the water. There are three pallet tanks (1, 2, and 3), which are filled with wastewater from a common pipe (8).

The setup shows:

Tank (1) is full, tank (2) has liquid in the overflow tray, and tank (3) is empty.

A valve (4) over each pallet tank is open if the pallet tank is to receive wastewater.

Only one pallet tank is filled at a time.

Each pallet tank has a sensor (5) mounted on the side to detect if it is filled with wastewater. The sensor is a Normally Closed (NC) contact.

When the pallet tank is full, valve (4) should close so that no more wastewater enters the pallet tank.

**Collection trays**

Under each pallet tank is a collection tray (6) used to collect wastewater if the pallet tank is leaking. The collection tray has a float switch (7), which is a Normally Closed (NC) contact. If the float switch detects liquid, the valve (4) should close so that no more wastewater enters the pallet tank.

**Signal tower**

Each pallet tank has a signal tower with three lamps, which means:

Light turned on when the valve is closed, and the pallet tank is filled.
Light turned on when the valve is open, and the pallet tank is being filled.
Light turned on when there is liquid in the collection tray.

You need to design your own control panel and define variable names.

**Exercise**

Write a PLC program that meets the above requirements.

 

## 11.7 Controlling multiple conveyor belts

In this exercise, you must design and write a PLC program for a small system consisting of three conveyor belts used for pallet transportation.

Illustration of the small system:

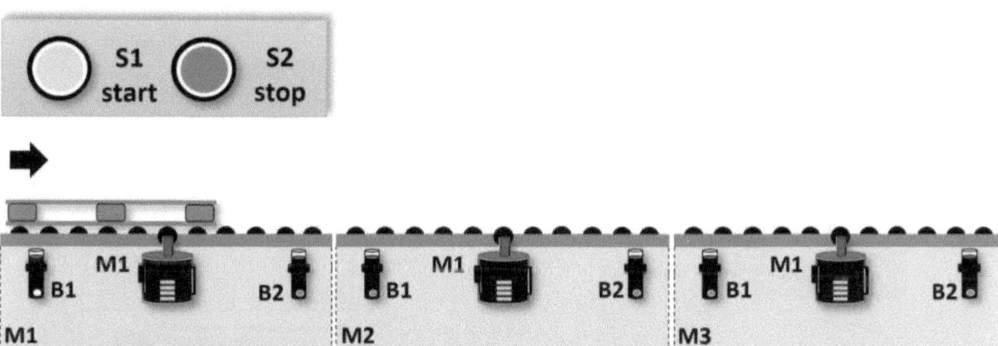


**Module-based conveyor system**

The three conveyor belts are constructed similarly, each equipped with a motor **M1** that drives the conveyor belt and two sensors **B1** and **B2**. The motor runs at a constant speed when there is a **TRUE** signal on **M1**. Each of the two sensors gives a **TRUE** signal when a pallet is directly above the sensor. The conveyor belts are all constructed identically to make the expansion of the entire system modular and to save costs.

**Energy-saving function**

To save energy, a conveyor belt should only operate when there is a pallet to move. This means that a conveyor belt should start only on a rising edge (**FALSE** to **TRUE** signal) from sensor **B2** from and the previous conveyor. And stop on a falling edge (**TRUE** to **FALSE** signal) from sensor **B1** on the next conveyor.

**Starting and stopping the system**

All three conveyor belts are controlled by a common PLC. The small system is started using the manual push button **S1** and stopped using the stop button **S2**. When the stop button **S2** is pressed, all conveyor belts should stop. It should be possible to resume operation after pressing the start button **S1**, even if a pallet is located between sensor **B1** and **B2** on a conveyor belt.

**Exercise**

Design and write a PLC program to control the small system.

## 11.8 Programming a washing machine

In this exercise, you must design and write a PLC program for a washing machine. The washing machine is used for washing clothes.

To the right, a cross-section of the washing machine is shown:

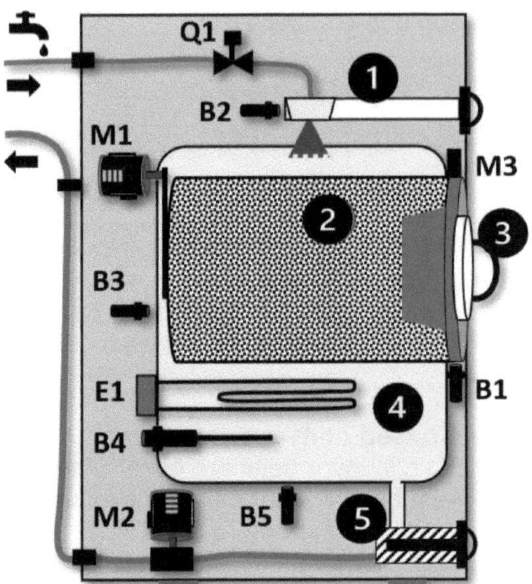


Explanation of the numbers:

| | |
|---|---|
| 1 | Soap drawer, which contains soap for one wash (soap dispenser box).<br>The soap drawer is pulled out when it needs to be refilled.<br>When sensor **B2** gives a **TRUE** signal, the soap drawer is closed correctly, and the washing machine can start.<br>When **Q1** gets a **TRUE** signal, the valve is open. This means that clean water can flow down through the soap drawer and flush the soap into the drum. |
| 2 | Drum where the dirty laundry is placed. |
| 3 | Door to the drum.<br>Sensor **B1** gives a **TRUE** signal when the door is closed.<br>Motor **M3** locks the door so that it cannot be opened when the washing machine is in operation.<br>Motor **M1** must not start if **B1** gives a **FALSE** signal (door open). |
| 4 | Area filled with water when the washing machine is washing.<br>When the valve **Q1** gets a **TRUE** signal, water enters the washing machine.<br>Level sensor **B5** gives a **TRUE** signal when there is no water in the drum.<br>Level sensor **B3** gives a **TRUE** signal when the drum is filled with water.<br>Motor **M2** pumps the dirty water out of the drum and into the sewer. |
| 5 | A cleaning filter ensures that large particles in the dirty water are retained and do not damage the pump **M2**. |

 

**Heating element**

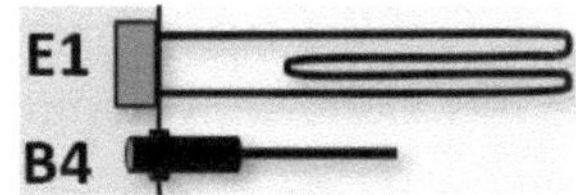


The water in the drum is heated with a heating element **E1**. A temperature sensor **B4** measures the temperature of the water. The heating element should only be turned on if sensor **B3** detects water (gives a **TRUE** signal), as the missing water can damage the heating element.

**Drum motor**

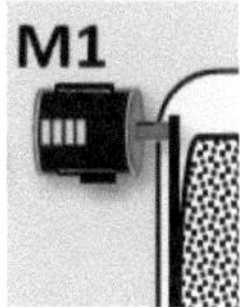


A motor **M1** rotates the washing drum, allowing the clothes inside the drum to be washed. The motor can operate at various speeds and in both forward and reverse directions, ensuring thorough cleaning and centrifugation (motor spins rapidly). Motor **M1** may only start if the door is closed.

**Wash programs**

The following wash programs should be available:

| Program | Name | Temperature | Max RPM | Time in minutes |
|---|---|---|---|---|
| 1 | Normal wash | 40 | 1400 | 120 |
| 2 | Boil wash | 60 | 1600 | 150 |
| 3 | Fine wash | 30 | 800 | 60 |
| 4 | Eco program | 40 | 1200 | 120 |
| 5 | Rinse | - | 1000 | 10 |
| 6 | Quick wash | 40 | 1600 | 30 |
| 7 | Pumping | - | - | - |

Where RPM stands for revolutions per minute.

**Alarm monitoring (alarm supervision)**

The following alarms should be present in the PLC program:

| Alarm | Text | Description |
|---|---|---|
| **A01** | No water | Valve **Q1** is open, but after some time, there is no **TRUE** signal from level sensor **B3**. |
| **A02** | Heating error | Heating element **E1** is turned on, but temperature sensor **B4** detects no temperature increase. |
| **A03** | Pump error or filter error | Pump motor **M2** is turned on, but after some time, level sensor **B5** still detects water. |
| **A04** | Sensor error | Level sensor **B3** detects water, but level sensor **B5** does not. |

You need to design and describe the control panel (HMI), in which the user can select wash programs and start and stop the washing machine. It should also be possible to view alarms on the control panel.

You need to define and describe the electrical signals for motor **M1**.

**Exercise**

Design a PLC program for the washing machine and then write it.

# 11.9 PLC program for an injection moulding machine

In this exercise, you must write a PLC program for an injection moulding machine:

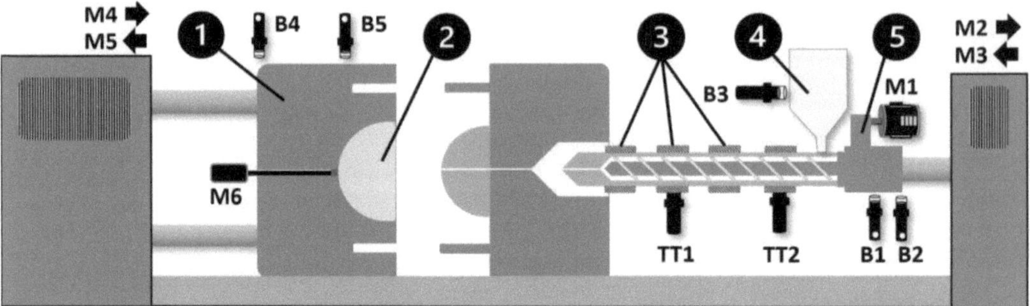


The machine is a simplified version of an injection moulding machine, as a real one often has more sensors and mechanical parts than shown in the image above.

**Plastic bowl**

The machine is used to mould a round plastic bowl like this:
(cross-section of the bowl)

To mould a plastic bowl, small plastic pieces (plastic granules) are needed, which are filled into the hopper ❹. Heating elements ❸ heat and melt the small plastic pieces, transforming the plastic into a liquid mass. The injection unit ❺ is used to inject the liquid plastic mass into the mould ❷.

**Operator panel**

The machine has an operator panel used to operate the machine, as shown to the right:

Pressing the start switch **S1** should start the machine.

Pressing the stop switch **S2** should stop the machine, and the following actions should be performed to return the machine to the starting position:

1) The mould ❶ is pushed to the left.
2) Motor **M1**, which controls the screw, is stopped.
3) The injection unit ❺ is pushed to the right.

 

Description of the machine:

| No. | Description |
|---|---|
| 1 | Movable mould:<br>The entire movable mould moves forward to the right when signal **M4** is **TRUE** and backward when **M5** is **TRUE**. **M4** and **M5** must not be **TRUE** at the same time.<br>If there is a **TRUE** signal from sensor **B4**, the mould is pushed all the way to the left.<br>If there is a **FALSE** signal from sensor **B5**, the mould is pushed all the way to the right, ready for a new moulding. |
| 2 | Mould that needs to be filled with liquid hot plastic.<br>Once the liquid plastic has cooled and solidified after 30 seconds, the plastic bowl can be released from the mould with an electric mandrel **M6**. |
| 3 | Electric heating element that heats and melts plastic granules.<br>Temperature is measured with two sensors, **TT1** (temperature transmitter) and **TT2**.<br>Both sensors must measure 200 degrees Celsius for the plastic to be melted and the moulding process to start. |
| 4 | Hopper for plastic granules. If there are not enough plastic granules in the hopper, sensor **B3** will give a **TRUE** signal, and the machine must stop. The machine must not be started until more plastic granules are added to the hopper. |
| 5 | The injection unit can be pushed forward towards the mould with motor **M3** and backward with motor **M2**.<br>A motor **M1** drives a screw auger, which presses the liquid plastic into the mould. |

Below are the steps required to mould a plastic bowl:

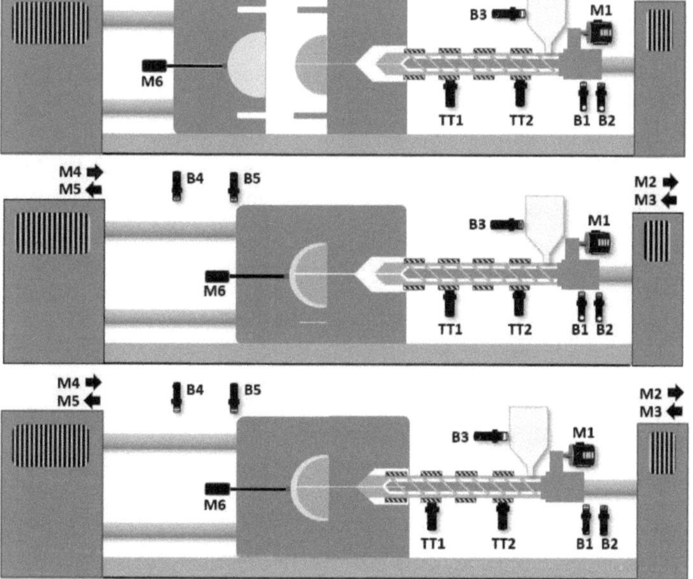


The machine is ready to start when the manual start switch **S1** on the operator panel is activated.

The movable mould should be pushed all the way to the right with motor **M4**, so the mould is completely closed.

The mould is in place when sensor **B5** gives a **FALSE** signal.

The injection unit is pushed into the mould with motor **M3**.

Sensor **B1** gives a **FALSE** signal when the injection unit is fully inserted.

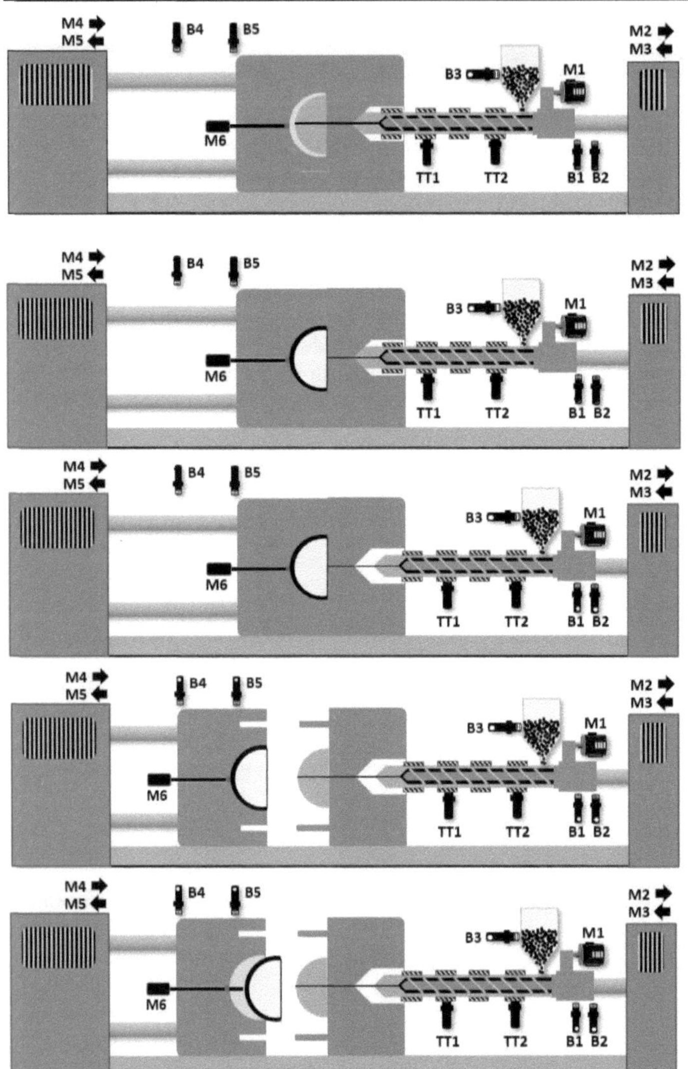


Plastic granules have now been added to the hopper.

The granules fall into the injection unit and are heated to melt them. The molten plastic is injected into the mould with a screw controlled by motor **M1**.

After 10 seconds, the mould is filled.

The screw controlled by motor **M1** is stopped.

The injection unit is retracted with **M2**.

Sensor **B2** gives a **TRUE** signal when the injection unit is fully retracted.

Once the plastic has solidified, the movable mould is moved to the left with motor **M5**.

Sensor **B4** gives a **TRUE** signal when the mould is fully moved to the left.

With an electric mandrel **M6**, the finished plastic bowl is pushed out of the mould.

**M6** must receive a **TRUE** signal for 2 seconds to push out the plastic bowl.

**Exercise**

Write a PLC program for the moulding machine.

 

# 11.10 Pump station with three pumps (alternating operation)

In this exercise, you must write a PLC program for a pump station with three pumps:

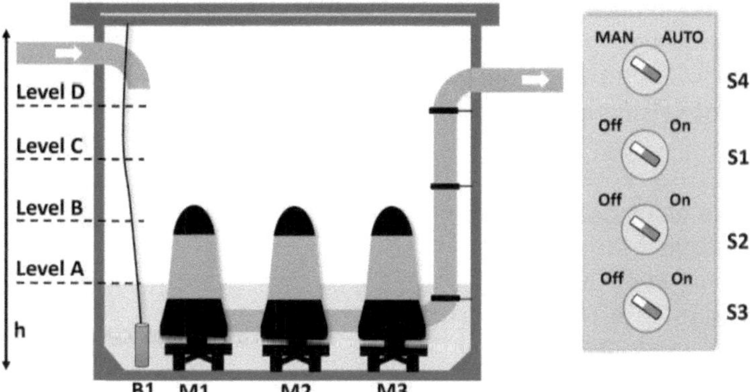


**Description**

The pump station contains three submerged pumps (**M1**, **M2**, **M3**), an analog level sensor **B1**, and an operating panel.
The height of the pump well is indicated by the height **h** and can be set to 6 meters.
The analog level sensor **B1** thus has a measurement range of 0 to 6 meters.
The control panel has a rotary switch **S4** for manual and automatic operation.

**Manual operation**

The pumps can be turned off or on individually with the rotary switches **S1**, **S2**, or **S3**.

**Automatic operation**

The rotary switch must be in the "on" position for the pump to be included in the automatic operation.

**Automatic mode operation**

Pumps should turn on and off depending on the level in the pump station. The level is measured with sensor **B1**. This means that:
If the level in the pump station is below **Level A**, all pumps should be off.
When the level in the pump station rises from **Level B** to **Level C** or from **Level C** to **Level D**, one more pump should be turned on. Thus, when the level in the pump station is above **Level D**, all pumps should be in operation.
When the level drops from **Level C** to **Level B** or from **Level B** to **Level A**, one pump should be turned off in order to avoid a pump stopping shortly after it has just started (this is called hysteresis).
In addition, an alternating operation must be used, so pumps should start and stop alternately when the level in the pump station changes. Furthermore, there should be 24 hours of alternating operation.

**Exercise**

Write a PLC program based on the description.

## 11.11 Development of PLC control for an elevator

A company develops and delivers elevators.

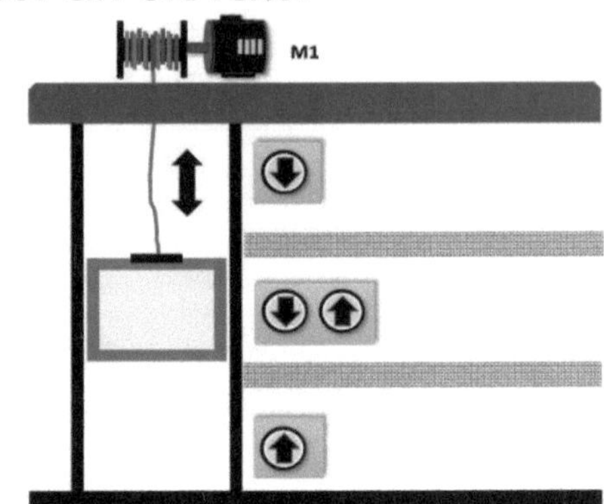


You have just been hired as a consultant and are tasked with assisting the company in developing a PLC control system for an elevator.

The elevator looks similar to the one shown to the right:

**Exercise**

1) Find suitable electrical components and sensors to control the elevator.
2) Write a control specification.
3) Write a PLC program for the elevator.

## 11.12 Development of a control system for mobile traffic lights

This exercise is to develop a PLC program for mobile traffic lights that can be used for roadworks.

Two sets of mobile traffic lights must always be used for roadworks.

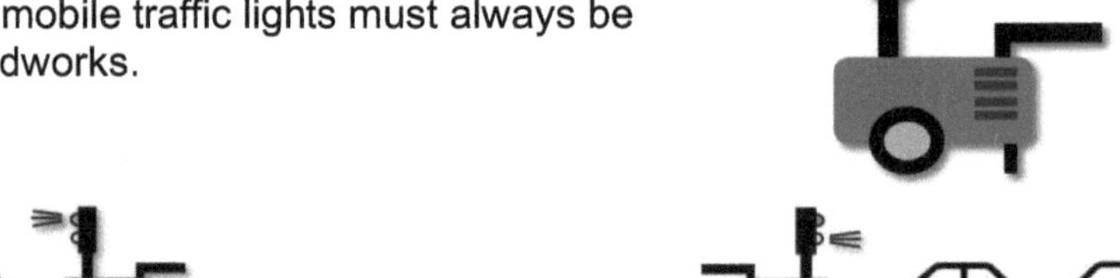

**Exercise**

Write a control description, which must also include a description of the electrical components you have chosen to use.

Since two mobile traffic lights are typically used together during roadworks, there should be a description of how the user synchronises the two traffic lights so that they do not have green lights at the same time.

The user should be able to adjust the duration time of the green/red light cycle, as the mobile traffic lights must be able to be used on variable road lengths.

You may choose an appropriate control method and data communication between the two mobile traffic lights.

Then write a PLC program that can be used for the two mobile traffic lights.

 

# 11.13 Palletizing with XY-robot controlled from a PLC

In this exercise, you must develop a PLC program for a system that stacks boxes onto a pallet.

The boxes come on a conveyor belt controlled by **M1**. When sensor **B1** gives a **TRUE** signal, the box should be moved onto the pallet. An XY-robot is used to move the boxes.

A total of 12 boxes need to be moved.

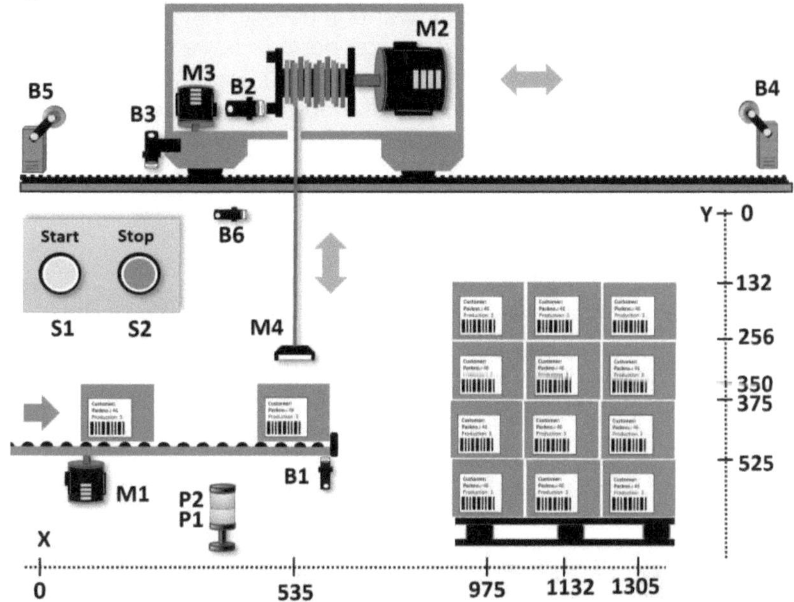


Description of the components:

| Name | Component | Description |
|---|---|---|
| **S1** | Push button | Start contact (NO), start palletising of boxes. |
| **S2** | Push button | Stop palletizing (NC). |
| **B1** | Sensor | Gives a **TRUE** signal when a box is ready to be picked up. |
| **B2** | Sensor | Encoder on hoist. Positions are shown at the Y-axis. |
| **B3** | Sensor | Encoder for x-direction. Positions are shown at the X-axis. |
| **B4**, **B5** | Sensor | Mechanical limit switch (NC). |
| **B6** | Sensor | Home position for Y-axis positions. |
| **P1** | Lamp | Turned on when the system is in operation (moving boxes). |
| **P2** | Lamp | Turned on when the system is not in operation. |
| **M1** | Motor | Motor for conveyor belt, which delivers the boxes. |
| **M2** | Motor | Motor for the hoist. |
| **M3** | Motor | Motor driving hoist in x-direction. |
| **M4** | Motor | Motor valve. If **TRUE**, there is vacuum suction for lifting a box. |

An XY-robot is also called a gantry robot or a cartesian robot.

**Exercise**

Write a control specification and then write a PLC program.

## 11.14 Bag filling at the fish factory

The company Fish Factory Ltd. wants a machine that can pack fish into bags. The fish vary in size and weight. The bags need to have the same weight, so there will often not be the same number of fish in each bag. If there is too little weight in the bag, consumers will complain, and if there is too much weight in the bag, money is lost. The challenge is to design a machine and write a PLC program that can ensure the correct number of fish and weight in each bag.

Illustration of the machine:

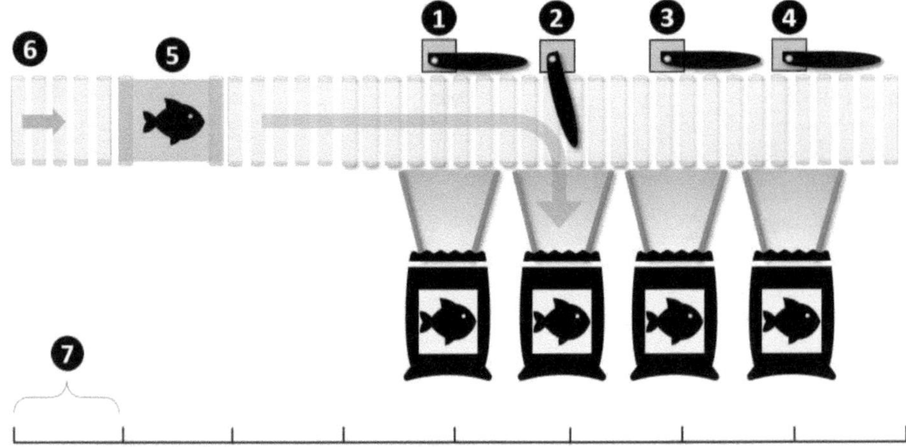


**Description**

The fish is placed on a conveyor belt at ❻.

Then it is transported to a scale at ❺, where it is weighed.

After the fish is weighed, it is further transported to an electrically movable arm at ❶❷❸❹; one of these arms pushes the fish into the bag.

The machine should signal an operator when a bag is full so the operator can remove the bag. The PLC program should be notified when a full bag has been removed and a new empty bag has been placed and is ready to receive fish.

The distance at ❼ is 0.5 meters.

The conveyor belt motor is controlled by a frequency converter.

It is assumed that there is only one fish on the conveyor belt at a time!

 

Selected data for the machine:

| Description | Data | Unit | Comments |
|---|---|---|---|
| Fish, size | 125 - 225 | [g] | |
| Weight, measurement range | 0 – 0.5 | [kg] | Measurement uncertainty +/- 1%.<br>Belt motor should stop during measurement. |
| Weight,<br>calibration | 24<br>2 | V<br>s | Digital signal input on scale.<br>Calibration time. |
| Belt motor<br>Ramp up time<br>Ramp down time<br>Roll diameter | 0 – 2000<br>500<br>500<br>30 | RPM<br>ms<br>ms<br>cm | Set on the frequency converter (VFD).<br>From stop to max. speed.<br>From max. speed to stop. |
| Bag size<br>Tolerance, min<br>Tolerance, max | 2.0<br>2.0<br>2.05 | Kg<br>Kg<br>Kg | |
| Belt speed | 4 | Km/h | |

**Exercise**

The following questions need to be answered:

1) How many fish can be expected in each bag?
2) What RPM (revolutions per minute) should be set on the frequency converter controlling the conveyor belt?
3) How long will it take (approximately in seconds) from placing a fish on the conveyor belt to it being in a bag?
4) How many bags can the machine pack in one hour?
5) How do you think the operator should be notified when a bag is filled?
6) How long should the fish be on the scale?
7) What types of signal inputs and outputs are needed for a PLC for the machine?
8) Draw flowcharts showing the steps a fish must go through to end up in a bag.
9) What type of PLC programming language would be best to use?
10) How is it possible to test the PLC program without using real fresh fish of differing weights?
11) Based on your answers to the questions above, design and write a PLC program for the machine including a user interface.
12) Create a factory acceptance test (FAT) document and test your program.

## 11.15 Controlling of autonomous mobile robots in a factory

In this exercise, you must develop a PLC program that can manage two mobile robots assigned to transport packages from worktables to the warehouse.

When an autonomous mobile robot (AMR) follows a specific or predetermined route, it is referred to as an Automated Guided Vehicle (AGV).

Overview of the factory floor:

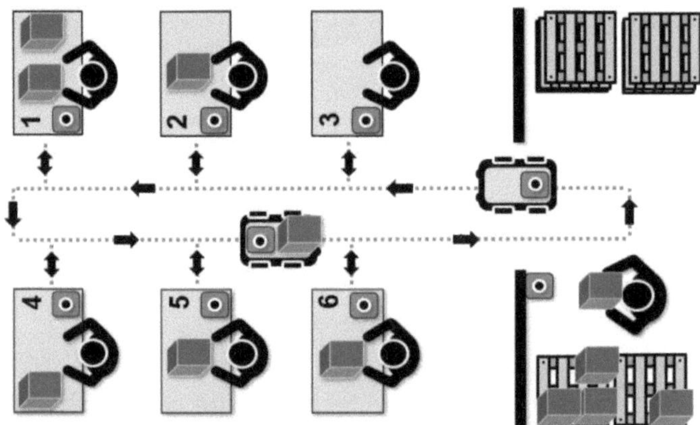


**Description**

There are six workbenches where employees place goods into packages. When a package is ready for pickup at a workbench, the employee presses a push button, which sends a signal to a PLC; this ensures that the package is picked up by the next available mobile robot. Once the PLC receives the message, the push button at the workbench turns the light on, indicating to the employee that the message has been received. The light should turn off again when the package has been picked up.

When a mobile robot arrives at the workbench, the employee places the package on it and presses a push button on the mobile robot, indicating to the robot that it should drive to the warehouse with the package.

At the warehouse, the mobile robot stops. A warehouse worker takes the package from the mobile robot and presses a push button on the wall, indicating to the mobile robot that it can take the next trip.

The two mobile robots follow a predetermined route, and if one mobile robot is stopped at a workbench, the other mobile robot can pass by.

**Assumptions for this exercise**

There can be up to 15 pickups in the queue.

A mobile robot takes 20 seconds in this exercise to travel from the warehouse and back to the warehouse with a package.

An employee can request multiple pickups shortly after each other.

**Exercise**

Write a PLC program that can manage the queue for picking up packages.

 

## 11.16 Simulate FIFO at a small factory

There is the following small factory consisting of four machines and conveyor belts:

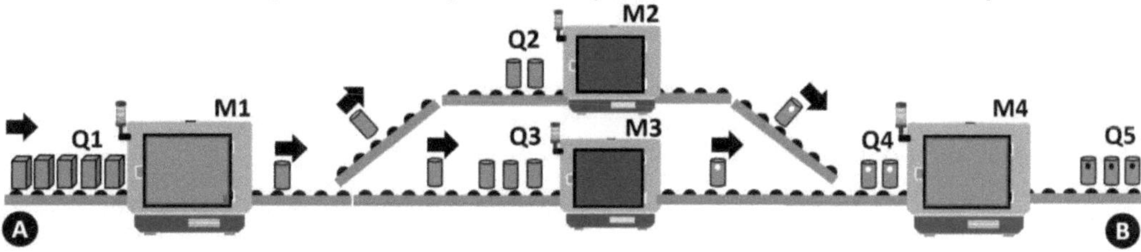


**Specification**

The machines **M1**, **M2**, **M3**, and **M4** process one item at a time.
Machine **M1** can process up to 15 items per minute.
Machine **M2** can process up to 6 items per minute.
Machine **M3** can process up to 8 items per minute.
Machine **M4** can process up to 12 items per minute.
Transport time for items between **M1** and **M2/M3** is 8 seconds.
Transport time for items between **M2/M3** and **M4** is 6 seconds.

**Description**

Items are placed on the conveyor belt by an operator at (point A). The items then enter a queue (**Q1**) in front of machine **M1**, which automatically takes a new item from the queue **Q1**. Machines **M3** and **M2** perform similar tasks. When items exit machine **M1**, they need to be distributed between machines **M2** and **M3**. You must decide on the optimal distribution method.
Finally, all items are fully processed by machine **M4**.

**Queue system**

Since the machines cannot process the same number of items per minute, a FIFO (first in first out) buffer (queue) is needed in front of each machine so that items can wait in line before entering it. After machine **M4**, there is also a queue **Q5**, as items are manually removed from the conveyor belt (point B). When an item is taken off, the operator must press an HMI (or a push button) to signal the PLC that the item has been removed from the conveyor belt and therefore the queue **Q5**.
Each queue can hold a maximum of five items.

**Production Number**

When an item is placed on the conveyor belt in front of machine **M1**, it must have a unique production number, which can either be entered via HMI (user interface) or generated automatically. This production number must accompany the items throughout the small factory.

**Exercise**

Program a PLC solution that simulates the entire small factory.

 

## 11.17 Data collection from machine to pivot table

In this exercise, you must collect data from a machine and display it in a pivot table.
The machine is shown on the right:

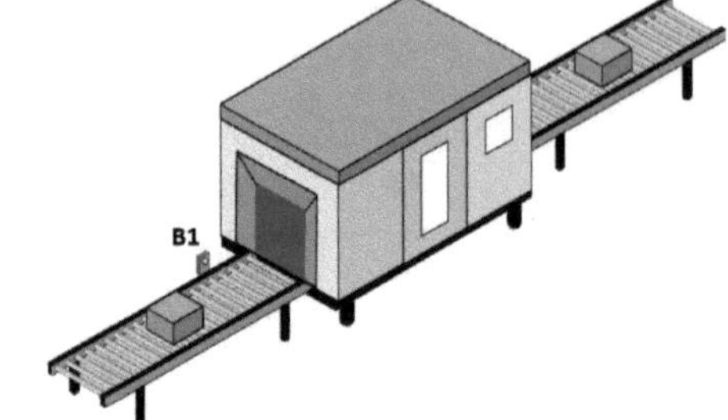


### Operation

A sensor **B1** is used to detect a package that is entering the machine.

### HMI

Displays a pivot table (column chart) with five columns. Each column shows the number of packages that have passed sensor **B1** within one minute. Every time one minute elapses, the columns shift to the left, so the column that is more than 5 minutes old is no longer displayed. The newest column is therefore to the right, and the column increases by one each time a package passes sensor **B1**. This column thus does not show the number of packages for a full minute but the ongoing period.

Illustration of the HMI:

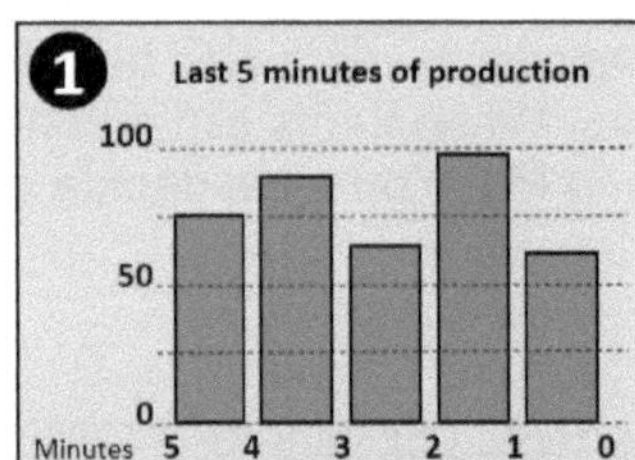


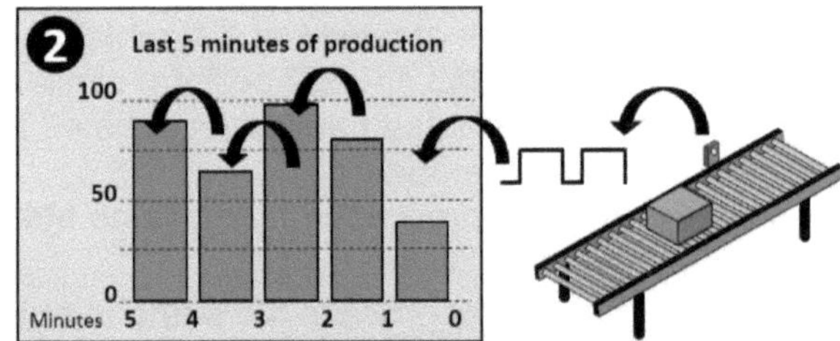


The two illustrations show the following:

- ❶ Displays the number of packages for the latest 5 minutes of production.
- ❷ Here, all columns from ❶ have been shifted one position to the left.
  Data from sensor **B1** is counted in the rightmost column.

### Data from PLC to HMI

The columns for the pivot table configured on the HMI are thus dynamic, so the height of the column is based on the value coming from the PLC.
The value in each column is in the range 0 to 100.
The five columns on the HMI are named **Col01**, **Col12**, **Col23**, **Col34**, and **Col45**.
The first column, **Col01,** is used to count the package, while the other four columns contain historical data.

### Exercise

Write a PLC program that collects data from **B1** and writes data to the five columns.

 

## 11.18 Data collection from three parking garages

In this exercise, you must design a solution and write a PLC program for collecting data from three parking garages.

Image of a typical parking garage:

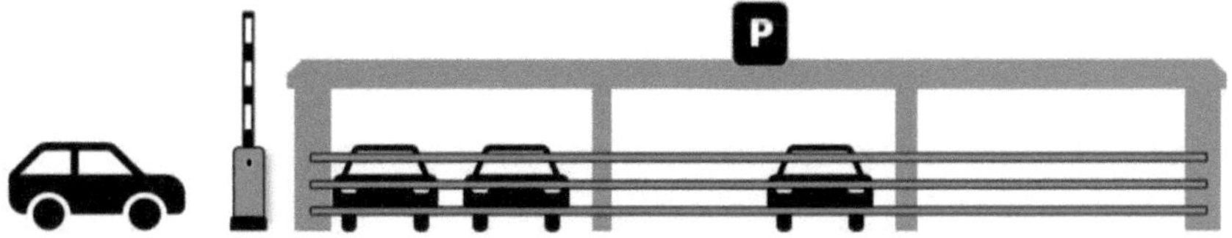


The parking garage is open every day from 6:00 AM to 8:00 PM. (06:00 to 20:00)
Each parking garage has its own PLC for local data collection.
There is a sensor at the entrance and exit, which registers when a car passes.

**Data collection requirements for a parking garage:**
For each hour, the following should be recorded:

A) The number of cars that entered the parking garage in the past hour.
B) The number of cars that exited the parking garage in the past hour.

For each day, the following should be calculated (once a day when the parking garage is closed):

C) The average number of cars that entered the parking garage per hour.
D) The total number of cars that entered the parking garage daily.

The PLC should store data for the past two days.

**Data collection from the parking garages:**
Every day, data from points C) and D) should be sent to a shared database installed on a common PC. Data is transmitted using a comma-separated ASCII-file (a flat text file). The PLC saves data in an ASCII-file, which is then loaded into a shared database.

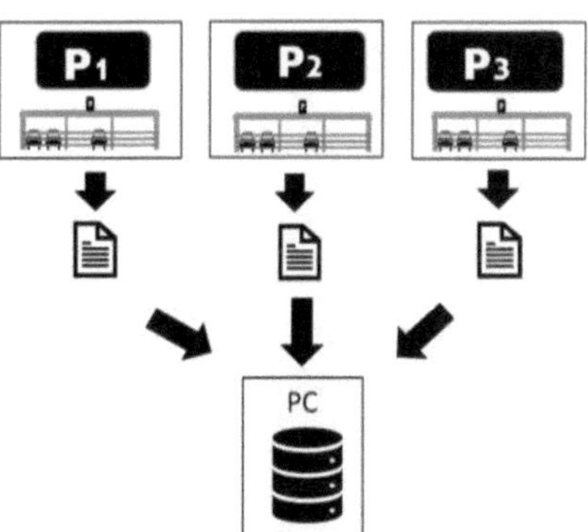


**Exercise**

1) Develop a program design proposal for each PLC in each parking garage. The proposal should include the necessary **ARRAYs** and a description of how data is collected and stored.
2) Develop a design proposal for the ASCII-file used for data collection.
3) Write a PLC program for each of the three parking garages.

# 11.19 Powder mix process with batch control

This exercise is to write a PLC program for a plant that mixes water with powder.

The plant consists of a large round mixing tank and four small tanks (**C1** to **C4**), each of which can contain various types of powder.

Illustration of the plant:

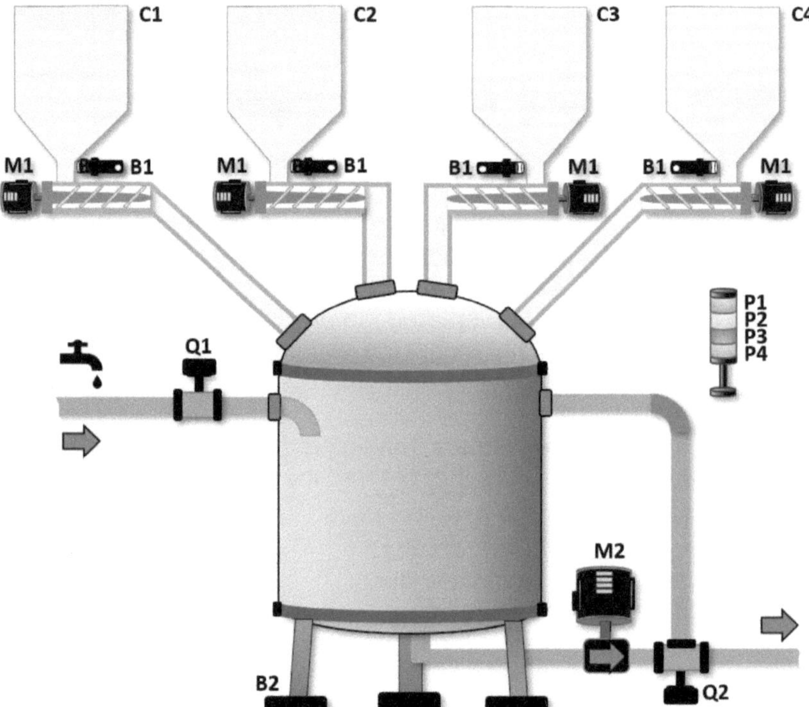


### Dosing and calibration

Below is a tank with powder:

Dosing with powder is calibrated. This is done by turning on motor **M1** for 1 second (**TRUE** signal to the motor) and weighing the amount of powder that has been dosed.

The four tanks are calibrated and provide the following amount of powder per second:

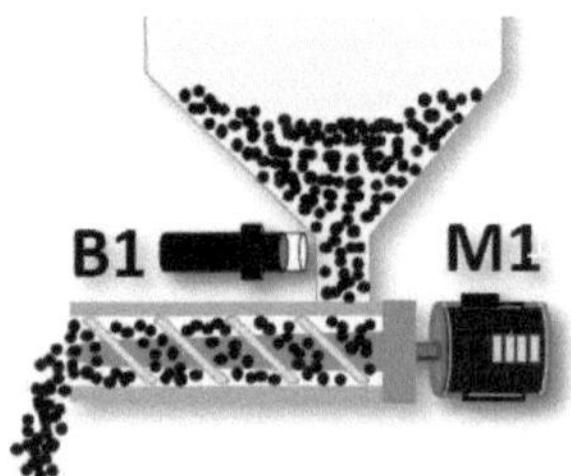


| Tank | Weight | Unit |
|---|---|---|
| C1 | 11 | gram |
| C2 | 9 | gram |
| C3 | 13 | gram |
| C4 | 12 | gram |

 

Description of the components:

| Name | component | Description |
|---|---|---|
| **C1-C4** | Tank | The tanks contain differing types of powder. |
| **M1** | Motor | Motor for dosing the powder. The motor is connected to an auger, which ensures a smooth dosing of powder when the motor is set to a **TRUE** signal. |
| **M2** | Pump | Motor with pump. When set to a **TRUE** signal, the pump operates, and its function depends on valve **Q2**:<br>When valve **Q2** has a **TRUE** signal, the contents of the tank are mixed. The liquid circulates from the bottom of the tank to the top, ensuring thorough mixing of powder and liquid.<br>When valve **Q2** has a **FALSE** signal, the tank is emptied. |
| **B1** | Sensor | Sensor gives a **TRUE** signal if a powder tank is empty.<br>The plant must not start if powder is needed from an empty powder tank. |
| **B2** | Load cell | The tank is placed on top of a load cell. Therefore, both the tank and its contents are weighed.<br>The load cell provides an analog 4–20 mA signal.<br>Measurement range: 0 to 3000 kg.<br>To calibrate the weight, the read value from the load cell is set to 0 in the PLC program when the tank is empty. |
| **Q1** | Valve | Dosing of water.<br>When set to a **TRUE** signal, the valve is open.<br>The amount of water needed for the mixture is measured with sensor **B2**. |
| **Q2** | Valve | Three-way valve.<br>When set to **TRUE**, the contents of the tank can be mixed.<br>When set to **FALSE**, the tank can be emptied. |
| **P1** | Lamp | Should be turned on if the mixing tank is empty and ready for a new mixture. |
| **P2** | Lamp | Should be turned on if motor **M2** is in operation and **Q2** is **TRUE**. Contents in the tank are mixed. |
| **P3** | Lamp | Should be turned on if motor **M2** is in operation and **Q2** is **FALSE**. The tank is emptied of its contents. |
| **P4** | Lamp | Should be turned on if a powder tank (**C1–C4**) is empty. |

**Naming**

The naming of tanks with powder follows the ISA-88 standard and constitutes a unit. Therefore, the components **B1** and **M1** can be annotated with the unit name (refer to ISA-88) (**C1** to **C4**) in front: For example, **C1_B1** and **C1_M1**.
The entire system constitutes a process cell.

**Table of batch quantities**

A series of standard mixtures with water and powder has been preconfigured.
The table below shows the various standard mixtures that should be selectable on the user panel (HMI):

| Batch no. | Name | C1 | C2 | C3 | C4 | Water | Mix time |
|---|---|---|---|---|---|---|---|
| 1 | CIP | 0 | 0 | 0 | 0 | 200 | 120 |
| 2 | Standard mix | 250 | 50 | 50 | 100 | 1000 | 300 |
| 3 | High C1 | 1000 | 50 | 50 | 100 | 1000 | 350 |
| 4 | Standard small | 25 | 5 | 5 | 10 | 100 | 100 |
| 5 | C3-C4 mix | 0 | 0 | 100 | 200 | 500 | 100 |
| 6 | Strong | 500 | 500 | 450 | 600 | 1000 | 500 |

Batch program number 1 is named CIP, which stands for clean-in-place and is a cleaning program for the tank and pipes.

**Units for quantities**

The units for dosing from **C1** to **C4** are in grams [g].
The unit for water is in litres [l].
The unit for mixing time is in seconds [s].
In this exercise, it is assumed that 1 litre of water weighs 1 kg.

**Requirements for the user panel and HMI**

Below are the requirements for the PLC program and user panel:

1) The user should be able to enter text describing the contents of each powder tank.
2) There should be an option for automatic operation, where batch no. 1 to 6 can be selected.
3) There should be an option for manual operation, where components can be operated individually.
4) It should be possible to start and stop the system.

**Exercise**

Based on the mentioned requirements and descriptions:

A) Design a user panel and HMI.
B) Design a program, and program the solution in a PLC.

**Extra exercise**

As an additional task, the user should have the option to delete or add new batches.

 